A Concise Introduction to Elastic Solids

Elastic Response

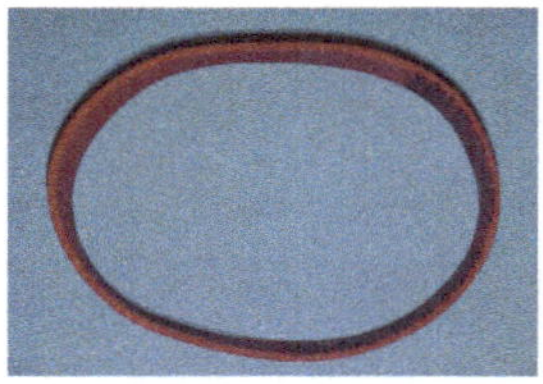

Elastic band before stretching

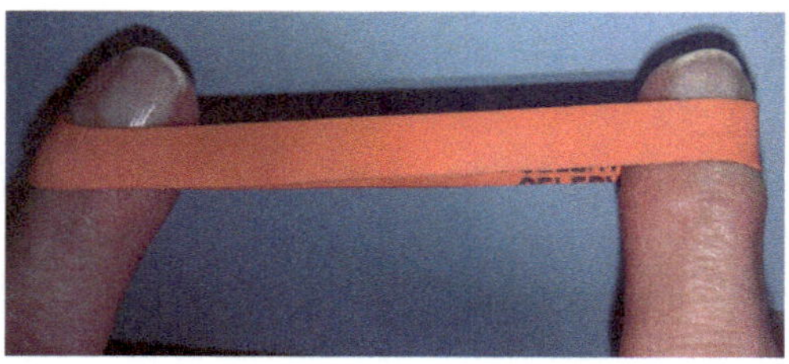

Stretched elastic band

Carl T. Herakovich

A Concise Introduction to Elastic Solids

An Overview of the Mechanics of Elastic Materials and Structures

Springer

Carl T. Herakovich
University of Virginia
Charlottesville, VA
USA

ISBN 978-3-319-45601-0 ISBN 978-3-319-45602-7 (eBook)
DOI 10.1007/978-3-319-45602-7

Library of Congress Control Number: 2016949614

Printed on acid-free paper

This Springer imprint is published by Springer Nature
The registered company is Springer International Publishing AG
The registered company address is: Gewerbestrasse 11, 6330 Cham, Switzerland

Dedicated to my parents:
Julia Marie (nee Buckley) Herakovich
John Bernard Herakovich
for setting an outstanding example

Preface

This treatise provides a broad overview of the definitions of fundamental quantities and the methods of analysis employed when solid materials that behave in an elastic manner are used in structural components. The presentation is limited to the linear elastic range of material behavior where there is a one-to-one relationship between load and displacement. Fundamental methods of analysis and typical results for structures made of elastic solid materials subjected to axial, bending, torsion, thermal, and internal pressure loading are presented. Stability of rods, analysis of plates, concepts of the finite element method, and mechanics of fibrous composite materials also are reviewed.

The presentation is at an introductory level suitable for advanced students in STEM (Science, Technology, Engineering, and Mathematics); it also can serve as an introduction or review for students and professionals in science and engineering. The treatise can be an introduction for specialists in fields such as aerospace engineering, mechanical engineering, materials science, and civil engineering. It is also intended to serve as an overview and possibly the only formal study of the subject for specialists in other fields of engineering and science. For a course of study at the college or university level, it is expected that it would, at most, be equivalent to a one-hour semester course. Finally, it is intended as an introductory overview for those in secondary science education and those teaching at the secondary level, as well as an introductory course for those studying in the arts and sciences at universities. As the emphasis on STEM education in the USA has increased in recent years, this treatise is a contribution to that effort. The treatise is written from the perspective of an engineer, one with more than fifty years of experience as an engineering professor. Introductory level exercises and their solutions are included.

This book is an expanded and improved version of a book entitled, Elastic Solids, previously self-published and no longer available.

Charlottesville, VA, USA
July 2016

Carl T. Herakovich

Acknowledgements

The author expresses his gratitude to Professor Michael Hyer of Virginia Tech and Professor James Simmonds of the University of Virginia for reviewing and making helpful suggestions to the earlier version, *Elastic Solids*. Thanks are also extended to my wife Marlene for over fifty-six years of support and encouragement. Elastic response photographs are by Carl and fingers by Marlene. And thanks to Michael Luby and Brian Halm of Springer for good advice.

Contents

About the Author

Carl T. Herakovich is (the Henry L. Kinnier) Professor Emeritus in the School of Engineering and Applied Science at the University of Virginia. He received a BS in civil engineering at the Rose Polytechnic Institute (now the Rose-Hulman Institute of Technology), MS in mechanics at the University of Kansas, and Ph.D. in mechanics at the Illinois Institute of Technology. He has taught mechanics courses at the University of Kansas, the Rose-Hulman Institute of Technology, the Illinois Institute of Technology, and in the Engineering Science and Mechanics Department at Virginia Tech for 20 years prior to joining the faculty at the University of Virginia in 1987. He was the founder of the NASA-Virginia Tech Composites Program, Chair of the Applied Mechanics Division, Vice President of the Basic Engineering Group of ASME, President and Secretary of the Society of Engineering Science, Secretary of the US National Committee on Theoretical and Applied Mechanics, and a US Delegate to the International Union of Theoretical and Applied Mechanics. He is the author of *Mechanics of Fibrous Composites* (John Wiley & Sons, 1998) and *Mechanics IUTAM USNC/TAM* (Springer, 2016) and author or coauthor of more than 135 scholarly papers. His e-mail address is herak@virginia.edu.

(November 15, 1958) Carl "Rocky" Herakovich (#22) presented game ball to Coach Phil Brown upon completion of perfect 8-0 season and 15 wins in a row over 2 years by the Rose Poly football team. Herakovich led the nation in scoring with 168 points in 8 games in 1958. That set a record for the most points scored in a season by a collegian in the State of Indiana and is still the record as of this writing some 58 plus years ago (Photo by Bob Crisp)

Chapter 1
Mechanics of Solids

1.1 Fundamentals

Mechanics of solids is concerned with understanding the response of solid materials (as opposed to fluids) when subjected to external forces and changes in environmental conditions such as temperature. The goal of research in the field over several centuries has been to develop uniformly agreed upon definitions of the fundamental physical quantities of interest and to express the interaction between these quantities as mathematical relationships. This treatise is concerned with solids whose properties are considered continuous over the region of interest. Thus, the term *continuum mechanics* has been coined to define the field of study. The properties may be uniform over the region of interest or vary in a predefined, continuous manner. Heterogeneous materials such as fibrous composite materials fall into the latter category.

A detailed history of solids mechanics for the years prior to 1952 was provided by Timoshenko (1953). There have been significant advancements in the field since the publication of Timoshenko's history book. In particular, the advent of fibrous composite materials and the development of modern computational methods have resulted in major advances in the field of solid and structural mechanics during the latter half of the twentieth century and the early twenty-first century. A recent book by Herakovich (2016) provides a survey of contributions in mechanics from antiquity to the present time (Fig. 1.1).

For the period of time covered by Timoshenko's history book, the solids of interest were primarily *isotropic*, that is, their properties were the same in every direction. Exceptions to this were wood and some crystal formations which exhibit *anisotropic properties*, (i.e., properties that vary with direction). More recently, solid mechanics is being applied to materials that exhibit anisotropic properties. The extensive use of fibrous composite materials in structures and studies in biomechanics has resulted in the necessity to routinely consider solids with anisotropic properties.

C.T. Herakovich, *A Concise Introduction to Elastic Solids*,
DOI 10.1007/978-3-319-45602-7_1

Fig. 1.1 Stephen Timoshenko

Another major shift since Timoshenko wrote his history book are the developments in computational methods such as the finite element method (Oden 2011). These methods were developed primarily by researchers working in solid mechanics and are now being used extensively in many areas of science and engineering.

1.2 Early History

The use of solid materials for structural applications dates to antiquity. The earliest uses of fabricated solids were for the implements and weapons fashioned during the stone, bronze, and iron ages. These devices were made through trial and error as opposed to mathematical representation of the physical object under consideration. The people who made these objects were artisans as opposed to *mechanicians*. The first application of mathematics to a structure may have been that by Archimedes (287–212 BC). He provided the mathematical bases for equilibrium of specific conditions of solid bodies under load.

It was not until Leonardo da Vinci (1452–1519) that mathematics was used in a more general manner to provide formalism to the study of mechanics of solids.

While da Vinci is best known for his painting *Mona Lisa*, he also used mathematics to study fundamental problems in solid mechanics. His *Proportions of Man* (Fig. 1.3) is widely used today in many illustrations. In one of his notes, da Vinci made the statement, "Mechanics is the paradise of mathematical science because here we come to the fruits of mathematics." (see Timoshenko (1953)) (Fig. 1.2).

Fig. 1.2 Leonardo da Vinci

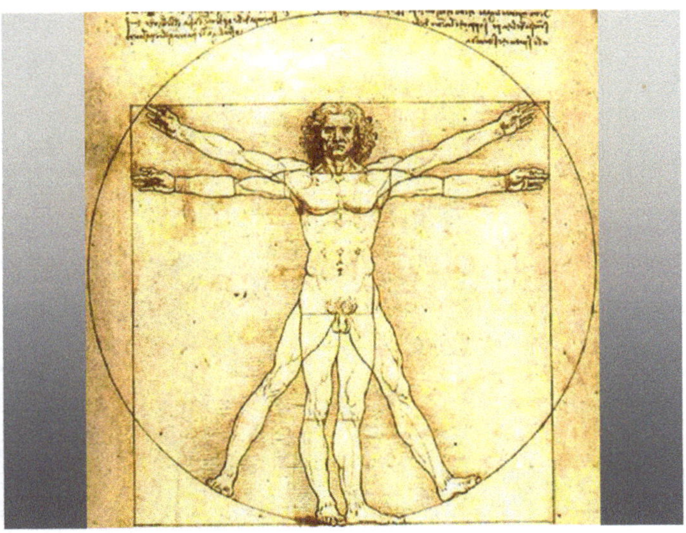

Fig. 1.3 da Vinci's proportions of man

After da Vinci, the next major step forward in the application of mathematics to solid mechanics is attributed to Galileo Galilei. While Galileo may be most famous for his work in astronomy, his book, *Two New Sciences* (1638), summarizes his work in mechanics and includes a section concerned with the mechanical properties of structural materials. Timoshenko considered this publication to be the birth of study in mechanics of solids.

Robert Hooke (1635–1703) is credited with proposing the linear relationship between force and deformation, a proposal based upon his own experiments. This linear relationship for the response of *elastic materials* is now referred to as *Hooke's law*. It is valid for responses which exhibit a one-to-one relationship between the applied force and the deformation, and where there is no permanent deformation upon unloading. Modern usage of this law is stated in terms of *stress* (force per unit area) and *strain* (deformation per unit length) (Fig. 1.4).

The study of the elastic response of solids is now referred to as *the mathematical theory of elasticity*. The theory was developed primarily during the nineteenth century by French mathematicians at l'École Polytechnique in Paris. The prime movers in the development of this theory were as follows: Navier (1785–1836), Cauchy (1789–1857), Poisson (1781–1840), and Saint Venant (1797–1886). The mathematical theory of elasticity provides the equations necessary to determine the stresses and strains in a body made of elastic solids when subjected to external forces or changes in environmental conditions.

Fig. 1.4 Galileo Galilei

Fig. 1.5 Augustin-Louis Cauchy

The rudiments of this theory are described in the following chapters through application to very specific problems of one-dimensional, two-dimensional, or three-dimensional configurations. All displacements are considered to be small and the solid bodies are in a state of static equilibrium. The presentation of fundamental problems is given for isotropic materials. However, the basic equations for laminated anisotropic materials and the complexities that they introduce also are provided and discussed. The presentation includes an introduction to modern methods of computational mechanics, namely the finite element method, for analysis of complex structural and material configurations (Figs. 1.5, 1.6, and 1.7).

Fig. 1.6 Siméon Denis Poisson

Fig. 1.7 Barré de Saint-Venant

References

Galilei, G. (1638). *Discourses and mathematical demonstrations relating to two new sciences*. Amsterdam: House of Elzevir.

Herakovich, C. T. (2016). *Mechanics IUTAM USNC/TAM*. Berlin: Springer.

Oden, J. T. (2011). *An introduction to mathematical modelling*. London: Wiley.

Timoshenko, S. P. (1953). History of strength of materials. McGraw-Hill Book Company, New York. Also Dover Publications, 1953.

Chapter 2
Stress

2.1 Cauchy Stress

The concept of stress was conceptualized best by Cauchy (1789) and put in final form by Saint-Venant (1797). Consider an infinitesimal cubical element (Fig. 2.1) that is in a state of static equilibrium. The forces per unit area on each face of the element are the *stresses* on that face. There are two types of stresses; normal stresses are perpendicular to the face of the element and shear stresses are parallel to the element face. Different notations have been used to denote stresses. In this treatise, σ will be used to denote normal stresses, and either σ or τ will be used to denote shear stress. The notation is further augmented with subscripts indicating the face upon which the stress is acting and the direction of the stress. Thus, σ_{xx} is the normal stress in the x-direction acting upon a face whose normal is in the x-direction. The shear stresses on the x-face are denoted τ_{xy} and τ_{xz} for shear stresses acting in the y- and z-directions, respectively.

Normal stresses are considered positive when tensile, and shear stresses are considered positive when the force vector is in the positive direction of the corresponding axis. All stresses shown in Fig. 2.1 are positive. It is noted that the stresses on a face correspond to the three components of an arbitrary force vector (per unit area) acting on the face. Further, the normal and shear stresses on opposite faces (not shown in Fig. 2.1) must be in opposite directions in order to maintain force and moment equilibrium. There are nine unique components of stress at a point.

C.T. Herakovich, *A Concise Introduction to Elastic Solids*,
DOI 10.1007/978-3-319-45602-7_2

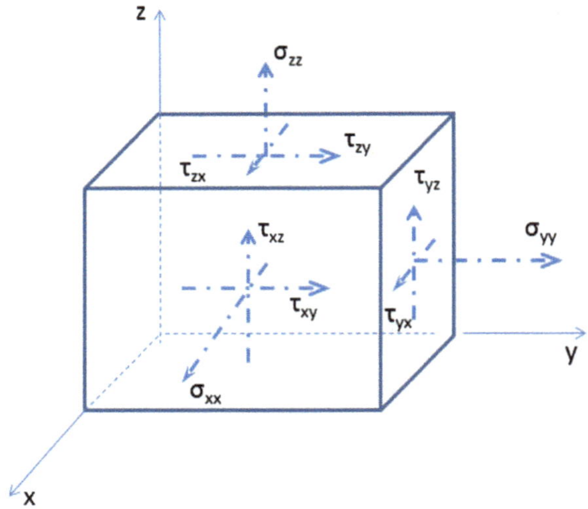

Fig. 2.1 Stress components

2.2 Plane Stress

A complete representation of the two-dimensional state of stress (plane stress with all z-components of stress = 0) at a point is depicted in Fig. 2.2. Note that the shear stresses on the three sets of opposite faces must be equivalent to couples that are in moment equilibrium. This establishes the condition that the shear stresses at a point are of equal magnitude, namely:

$$\tau_{xy} = \tau_{yx} \tag{2.1}$$

For the conditions of plane stress to exist in an x–y plane, it is understood that the z-components of stress are zero, i.e., $\sigma_{zz} = \tau_{zx} = \tau_{zy} = 0$.

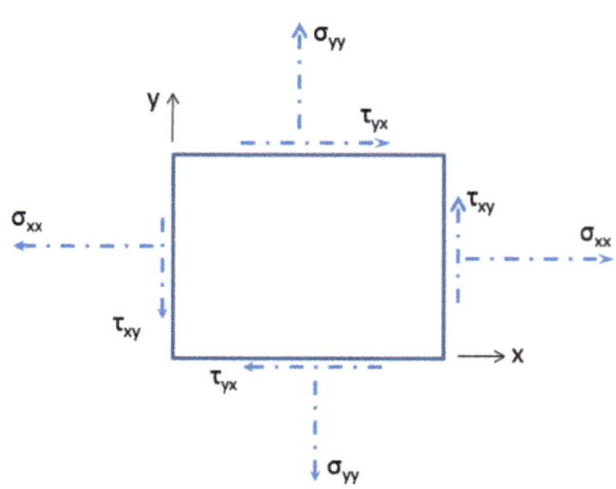

Fig. 2.2 Plane stress at a point

2.3 Stress Transformation

The plane state of stress at any point is a function of the plane of interest through that point. Ideally, it is desired to express the state of stress on any arbitrary plane passing through the point. This can be accomplished through consideration of the equilibrium equations for an arbitrary section passing through the element in Fig. 2.2. The resulting equations are called *stress transformation equations*.

Consider the free body diagram of a section of an element as shown in Fig. 2.3.

Force equilibrium in the x′- and y′-directions results in the following plane stress transformation equations.

$$\sigma_{x'x'} = \sigma_{xx}\cos^2\theta + \sigma_{yy}\sin^2\theta + 2\tau_{xy}\sin\theta\cos\theta \tag{2.2}$$

$$\tau_{x'y'} = -\left(\sigma_{xx} - \sigma_{yy}\right)\sin\theta\cos\theta + \tau_{xy}\left(\cos^2\theta - \sin^2\theta\right) \tag{2.3}$$

Following along similar lines for equilibrium, the normal stress $\sigma_{y'y'}$ is:

$$\sigma_{y'y^i} = \sigma_{xx}\sin^2\theta + \sigma_{yy}\cos^2\theta - 2\tau_{xy}\sin\theta\cos\theta \tag{2.4}$$

These stress transformation equations can be expressed in a more convenient form through the use of trigonometric identities (see Appendix) with the results:

$$\sigma_{x'x'} = \frac{\sigma_{xx} + \sigma_{yy}}{2} + \frac{\sigma_{xx} - \sigma_{yy}}{2}\cos 2\theta + \tau_{xy}\sin 2\theta \tag{2.5}$$

$$\sigma_{y'y'} = \frac{\sigma_{xx} + \sigma_{yy}}{2} + \frac{\sigma_{xx} - \sigma_{yy}}{2}\cos 2\theta - \tau_{xy}\sin 2\theta \tag{2.6}$$

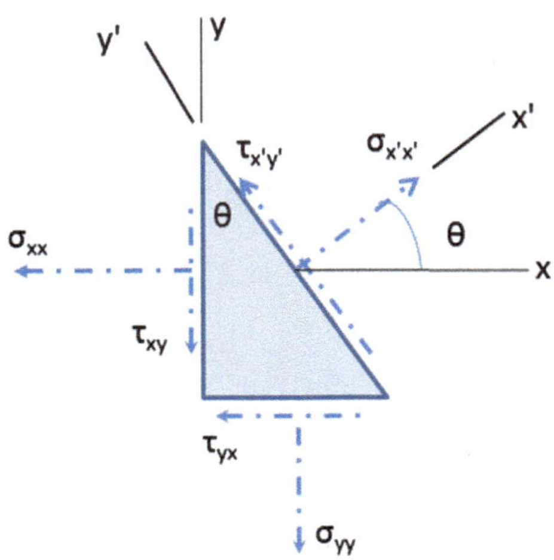

Fig. 2.3 Arbitrary section of plane stress element

$$\tau_{x'y'} = -\left(\frac{\sigma_{xx} - \sigma_{yy}}{2}\right)\sin 2\theta + \tau_{xy}\cos 2\theta \tag{2.7}$$

Defining $m = \cos\theta$ and $n = \sin\theta$, the transformation equations (2.2–2.4) can be written in matrix form as:

$$\begin{Bmatrix} \sigma_{x'x'} \\ \sigma_{y'y'} \\ \tau_{x'y'} \end{Bmatrix} = \begin{bmatrix} m^2 & n^2 & 2mn \\ n^2 & m^2 & -mn \\ -mn & mn & m^2 - n^2 \end{bmatrix} \begin{Bmatrix} \sigma_{xx} \\ \sigma_{yy} \\ \tau_{xy} \end{Bmatrix} \tag{2.8}$$

2.4 Principal Stresses

The normal stresses at a point vary with the orientation of the plane passing through the point as determined from equilibrium. The maximum and minimum values of the normal stresses are called the *principal stresses*. Taking the derivative of the normal stress equilibrium equation (2.4) with respect to the angle θ and equating to zero gives the angles, θ_P, corresponding to the principal planes. The resulting equation is:

$$\tan(2\theta_P) = \frac{2\tau_{xy}}{\sigma_{xx} - \sigma_{yy}} \tag{2.9}$$

The two angles θ_P from this equation differ by 90°. The planes defined by these angles are called the *principal planes*. One of the planes corresponds to the maximum normal stress and the other corresponds to the minimum normal stress. Combining these results (2.4 and 2.6) gives the expression for the principal stresses:

$$\sigma_{11,22}^{P} = \frac{\sigma_{xx} + \sigma_{yy}}{2} \pm \sqrt{\left(\frac{\sigma_{xx} - \sigma_{yy}}{2}\right)^2 + \tau_{xy}^2} \tag{2.10}$$

Using the (+) sign in this equation gives the maximum normal stress, $\sigma_{11}^{\max}$, and the (–) sign gives the minimum normal stress, $\sigma_{22}^{\min}$.

The maximum shear stress at a point may also be of interest. Setting the derivative with respect to θ of the shear stress equation (2.5) equal to zero gives

$$\tan(2\theta_s) = -\left(\frac{\sigma_{xx} - \sigma_{yy}}{2\tau_{xy}}\right) \tag{2.11}$$

The two solutions for θ_S correspond to the planes of maximum and minimum shear stress. They are equal in magnitude with the maximum shear stress being positive and the minimum shear stress being negative. The magnitude of the maximum shear stress is determined by combining the above equations with the result:

$$\tau_{\max} = \sqrt{\left(\frac{\sigma_{xx} - \sigma_{yy}}{2}\right)^2 + \tau_{xy}^2} \tag{2.12}$$

It can also be shown that the maximum shear stress can be expressed in terms of the principal stresses as:

$$\tau_{\max} = \frac{\sigma_{11} - \sigma_{22}}{2} \tag{2.13}$$

2.5 Pure Shear Stress

The condition of pure shear is a most interesting state of stress. As shown in Fig. 2.4, pure shear in an x–y plane is equivalent to positive and negative, pure, normal stresses, of the same magnitude as the shear stress, on the planes at angles 45° to the x–y axes. This explains why a brittle material such as a piece of chalk fails along a 45° plane when subjected to pure torsion (Chap. 7). The brittle material is relatively weak in tension and thus it fails along the plane that has the highest level of tensile, normal stresses.

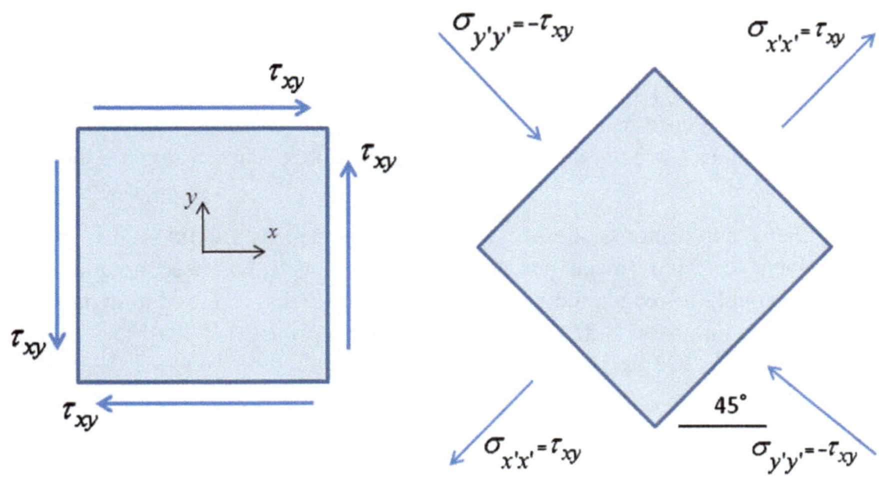

Fig. 2.4 Pure shear stress

2.6 Stress Tensor

The nine components of stress can be represented as a tensor quantity that obeys tensor transformation laws as described in the Appendix. The standard notation for the stress tensor is:

$$\sigma_{ij} = \begin{bmatrix} \sigma_{11} & \sigma_{12} & \sigma_{13} \\ \sigma_{21} & \sigma_{22} & \sigma_{23} \\ \sigma_{31} & \sigma_{32} & \sigma_{33} \end{bmatrix} \tag{2.14}$$

In (2.14), the indices 1, 2, 3 represent x, y, z in the rectangular Cartesian coordinate system. Further, we note that equilibrium requires that the stress tensor is symmetric with:

$$\sigma_{ij} = \sigma_{ji} \tag{2.15}$$

2.7 Units of Stress

Stress is the force per unit area. Thus, stress can be expressed as pounds per square inch (psi) or Pascals [Pa, which is defined as newtons per square meter in the International (SI) System of units]. The units on force are derived from Newton's second law of motion which states that F = m • a. One newton (N) is the force required to accelerate one kilogram (kg) of mass (m) at the rate (a) of one meter per second squared (m/s^2). Thus 1 N = 1 kg • 1 m/s^2. The unit of mass in the Imperial and United States customary systems is called a slug. It is the mass that accelerates at one ft/s^2 when a force of one pound (lb) is exerted on it. Thus, a slug has the units $1\ \text{slug} = 1\,\frac{\text{lb}\cdot \text{s}^2}{\text{ft}}$.

For many applications, stresses are often expressed as ksi [kips (10^3 lb) per square inch] or MPa (mega pascals, where mega = 10^6). These notations are employed simply to reduce the number of zeros (or digits) required to express the value of large numbers. Note that 1.0 MPa = 145.0 psi = 0.145 ksi.

2.8 Exercises

2.8.1 Confirm equation 2.2.
2.8.2 Confirm equation 2.5.
2.8.3 Confirm equation 2.7.
2.8.4 Confirm equation 2.9.
2.8.5 Plot the variation of normal stress on planes passing through a point if it is known that the state of stress is planar with $\sigma_{xx} = 40,\ \sigma_{yy} = -20,\ \tau_{xy} = 10.$

Appendix: Solutions

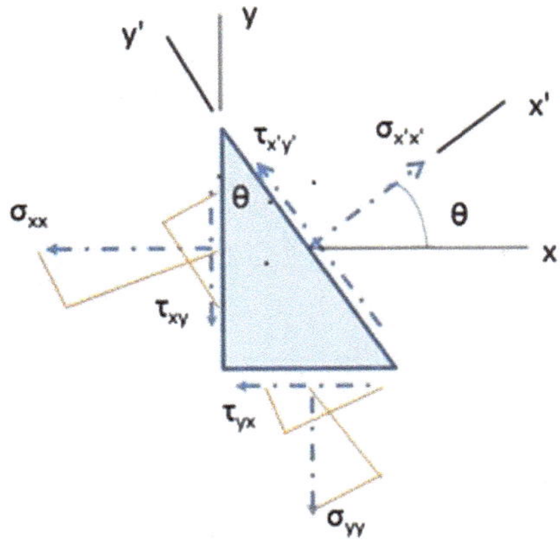

2.8.1 Confirm equation 2.2.

Solution
Equation (2.2) ⇒

$$\sigma_{x'x'} = \sigma_{xx}\cos^2\theta + \sigma_{yy}\sin^2\theta + 2\tau_{xy}\sin\theta\cos\theta$$

Summation of forces in the $x' - x'$-direction gives:

$$\sum F_{x'x'} = 0$$
$$\sigma_{x'x'}A = \sigma_{xx}\cos\theta A\cos\theta + \sigma_{yy}\sin\theta A\sin\theta + \tau_{xy}\cos\theta A\sin\theta + \tau_{xy}A\cos\theta\sin\theta$$
$$\sigma_{x'x'} = \sigma_{xx}\cos^2\theta + \sigma_{yy}\sin^2\theta + 2\tau_{xy}\cos\theta A\sin\theta$$

2.8.2 Confirm equation 2.5.

Equation (2.5) ⇒

$$\sigma_{x'x'} = \frac{\sigma_{xx} + \sigma_{yy}}{2} + \frac{\sigma_{xx} - \sigma_{yy}}{2}\cos 2\theta + \tau_{xy}\sin 2\theta$$

Solution
Using the trigonometric identities in (2.2):

$$\sin 2\theta = 2\sin\theta\cos\theta$$
$$\sin^2\theta = \frac{1-\cos 2\theta}{2}$$
$$\cos^2\theta = \frac{1+\cos 2\theta}{2}$$

$$\Rightarrow$$

$$\sigma_{x'x'} = \sigma_{xx}\frac{1+\cos 2\theta}{2} + \sigma_{yy}\frac{1-\cos 2\theta}{2} + \tau_{xy}\sin 2\theta$$
$$\sigma_{x'x'} = \frac{\sigma_{xx}+\sigma_{yy}}{2} + \frac{\sigma_{xx}-\sigma_{yy}}{2}\cos 2\theta + \tau_{xy}\sin 2\theta$$

2.8.3 Confirm equation 2.7.

Equation (2.7) $\Rightarrow$

$$\tau_{x'y'} = -\left(\frac{\sigma_{xx}-\sigma_{yy}}{2}\right)\sin 2\theta + \tau_{xy}\cos 2\theta$$

Solution
Summation of forces in the $y' - y'$-direction and using trigonometric identities gives:

$$\sum F_{y'y'} = 0 \quad \Rightarrow$$
$$\tau_{x'y'}A = A\cos\theta\left(\sigma_{xx}\sin\theta - \tau_{xy}\cos\theta\right) + A\sin\theta\left(\tau_{xy}\sin\theta - \sigma_{yy}\cos\theta\right)$$

$$\tau_{x'y'} = (\sigma_{xx}-\sigma_{yy})\sin\theta\cos\theta + \tau_{xy}\left(\sin^2\theta - \cos^2\theta\right)$$
$$\tau_{x'y'} = (\sigma_{xx}-\sigma_{yy})\frac{\sin 2\theta}{2} + \tau_{xy}\cos 2\theta$$

2.8.4 Confirm equation 2.9.

Equation (2.9) $\Rightarrow$

$$\tan(2\theta_P) = \frac{2\tau_{xy}}{\sigma_{xx}-\sigma_{yy}}$$

Solution

Normal stress on arbitrary plane is from (2.5) ⇒

$$\sigma_{x'x'} = \frac{\sigma_{xx} + \sigma_{yy}}{2} + \frac{\sigma_{xx} - \sigma_{yy}}{2}\cos 2\theta + \tau_{xy}\sin 2\theta$$

Setting derivative with respect to $\theta = 0$ for maximum and minimum ⇒

$$\frac{d\sigma_{x'x'}}{d\theta} = 0 = -2\sigma_{xx}\sin\theta\cos\theta + 2\sigma_{yy}\sin\theta\cos\theta + 2\tau_{xy}\left(\cos^2\theta - \sin^2\theta\right)$$

$$0 = \frac{-\sigma_{xx}\sin 2\theta}{2} + \frac{\sigma_{yy}\sin 2\theta}{2} + \tau_{xy}\left(\frac{1+\cos 2\theta}{2} - \frac{1-\cos 2\theta}{2}\right)$$

$$0 = \frac{\sin 2\theta}{2}\left(\sigma_{yy} - \sigma_{xx}\right) + \tau_{xy}\cos 2\theta$$

$$\tan 2\theta = \frac{2\tau_{xy}}{\sigma_{xx} - \sigma_{yy}}$$

2.8.5 Plot the variation of normal stress on planes passing through a point if it is known that the state of stress is planar with $\sigma_{xx} = 40,\ \sigma_{yy} = -20,\ \tau_{xy} = 10$

Solution

Plotting Eq. (2.2) ⇒

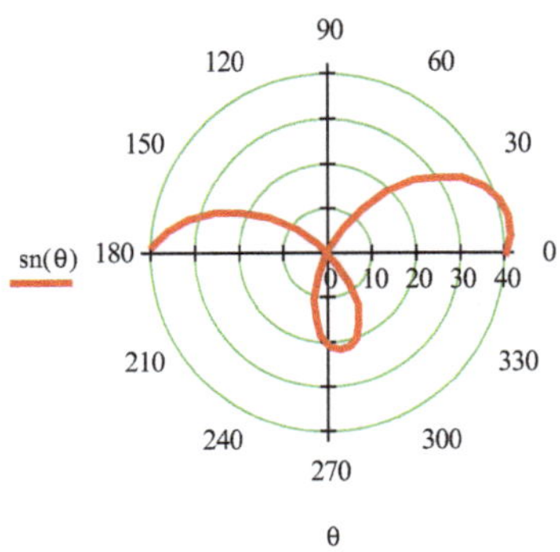

References

Cauchy (1789–1857) Cauchy, A. (1822). *Memoires de l'acdemie des sciences*, Paris, Vol. 7, (1827), pp. See, Bulletin de la societe philomathique, Paris (1823), p. 177.

Saint Venant (1797–1886) Saint-Venant, A. J. C. Barré de. (1845). Comptes Rendus, v20, p. 1765 and v21, p. 125.

Chapter 3
Strain

3.1 Normal Strain

Normal strain, ε, is the change in length, u, per unit length when a body, length L, is subjected to normal stress. For loading P in the x-direction (Fig. 3.1), we define the axial (normal) strain as:

$$\varepsilon_{xx} = \frac{u}{L} \tag{3.1}$$

For a three-dimensional cube, the normal strains are: $\varepsilon_{xx}, \varepsilon_{yy}$, and ε_{zz}. Positive normal stains are depicted in Fig. 3.2 which shows the original (solid) and deformed (dashed) shape of a cube. Since strain is defined as a length, u, divided by a length, L, it is a dimensionless quantity. However, it is often written as in/in, m/m, or %. strain.

3.2 Engineering Shear Strain

Engineering shear strain corresponds to the change in orientation of an element that is subjected to shear stress. More specifically, it is the total change in angle (expressed in radians) of an original right angle due to shear stress. Figure 3.3 depicts the (engineering) shear strain γ_{xy}.

C.T. Herakovich, *A Concise Introduction to Elastic Solids*,
DOI 10.1007/978-3-319-45602-7_3

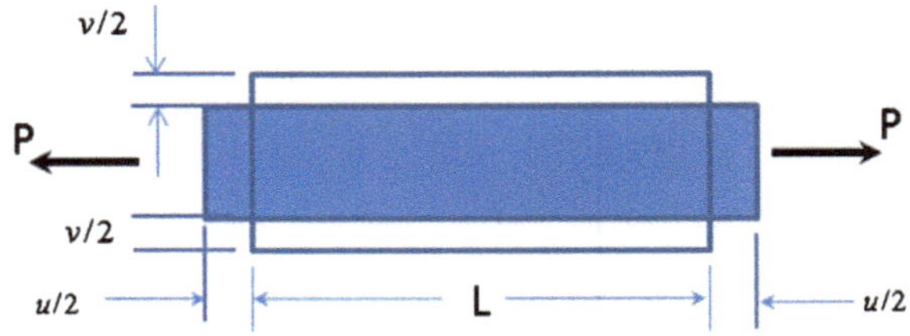

Fig. 3.1 Axial strains

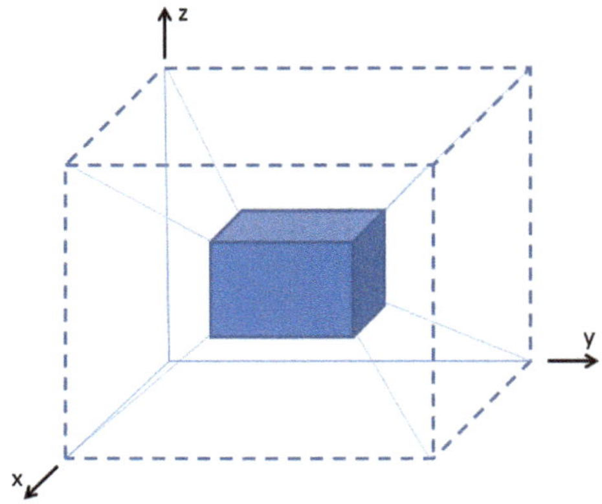

Fig. 3.2 Three-dimensional normal strain deformations

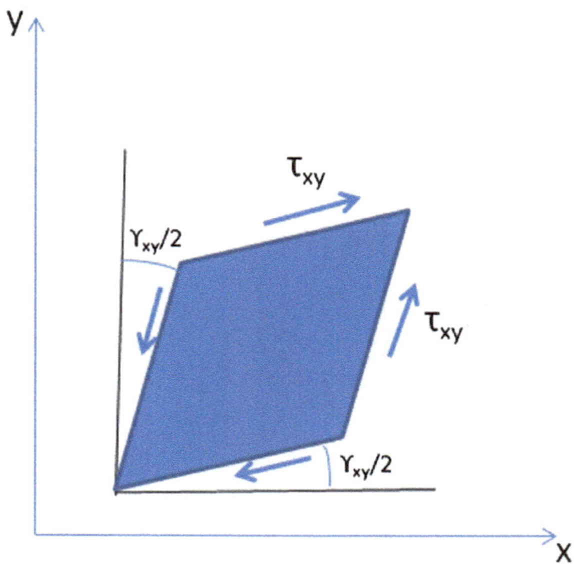

Fig. 3.3 Shear strains

3.3 Strain–Displacement Relationships

Normal strain and shear strain are defined in the preceding sections for specific cases of deformation. More generally, we can write strain–displacement relationships in terms of displacements $u, v, and\ w$ in the x-, y-, and z-directions, respectively, at a point in a solid. Normal strains are the rate of change of displacement in a specific direction. They can be expressed as partial derivatives (see Appendix for partial derivatives). The following strain–displacement equations are valid if the displacements are small and the first derivatives so small that squares and products of their partial derivatives are negligible.

For the three normal components of strain, we have:

$$\begin{aligned}\varepsilon_{xx} &= \frac{\partial u}{\partial x}\\ \varepsilon_{yy} &= \frac{\partial v}{\partial y}\\ \varepsilon_{zz} &= \frac{\partial w}{\partial z}\end{aligned} \tag{3.2}$$

The engineering shear strains correspond to the change of right angle in a plane and can be expressed in partial derivative form as:

$$\begin{aligned}\gamma_{xy} &= \left(\frac{\partial u}{\partial y} + \frac{\partial v}{\partial x}\right)\\ \gamma_{xz} &= \left(\frac{\partial u}{\partial z} + \frac{\partial w}{\partial x}\right)\\ \gamma_{yz} &= \left(\frac{\partial v}{\partial z} + \frac{\partial w}{\partial y}\right)\end{aligned} \tag{3.3}$$

3.4 Tensor Shear Strain

For transformation of shear strain, it is necessary to express the shear strain as a tensor quantity. Note that the engineering shear strain γ_{ij} is double the tensor shear strain ε_{ij} $(i \neq j)$. Thus, the tensor shear strains ε_{ij} are:

$$\varepsilon_{ij} = \frac{1}{2}\gamma_{ij} = \frac{1}{2}\left(\frac{\partial u_j}{\partial x_i} + \frac{\partial u_i}{\partial x_j}\right) \quad (i \neq j) \tag{3.4}$$

Note that strain is a symmetric tensor, i.e., $\varepsilon_{ij} = \varepsilon_{ji}$.

We can now write the general expression for all six components of the tensor strain, ε_{ij}:

$$\varepsilon_{ij} = \frac{1}{2}\left[\frac{\partial u_j}{\partial x_i} + \frac{\partial u_i}{\partial x_j}\right] \quad (i,j \Rightarrow x, y, z \ directions) \tag{3.5}$$

The tensor (3.5) is known as Cauchy's infinitesimal strain tensor.

3.5 Strain Transformation

It is often desired to express strains in an arbitrary x', y' coordinate system as was done for stresses (Chap. 2, Eqs. 2.4 and 2.5). The strain transformation equations for a rotation θ about the z-axis are:

$$\begin{aligned}
\varepsilon_{x'x'} &= \varepsilon_{xx}\cos^2\theta + \varepsilon_{yy}\sin^2\theta + \gamma_{xy}\sin\theta\cos\theta \\
\varepsilon_{y'y'} &= \varepsilon_{xx}\sin^2\theta + \varepsilon_{yy}\cos^2\theta - \gamma_{xy}\sin\theta\cos\theta \\
\gamma_{x'y'} &= -2(\varepsilon_{xx} - \varepsilon_{yy})\sin\theta\cos\theta + \gamma_{xy}(\cos^2\theta - \sin^2\theta)
\end{aligned} \tag{3.6}$$

Since strain is a second-order tensor quantity, using Appendix A.12, we can write the general transformation equation for rotation about the z-axis, with i, j = 1, 2, as:

$$\varepsilon'_{ij} = a_{1i}a_{1j}\varepsilon_{11} + a_{1i}a_{2j}\varepsilon_{12} + a_{2i}a_{1j}\varepsilon_{21} + a_{2i}a_{2j}\varepsilon_{22} \tag{3.7}$$

where

$$a_{ij} = \begin{pmatrix} \cos\theta & -\sin\theta \\ \sin\theta & \cos\theta \end{pmatrix} \tag{3.8}$$

Similar to the stress transformation equation (2.8), strain transformation can be written as:

$$\begin{pmatrix} \varepsilon_{x'x'} \\ \varepsilon_{y'y'} \\ \gamma_{x'y'} \end{pmatrix} = \begin{bmatrix} m^2 & n^2 & mn \\ n^2 & m^2 & -mn \\ -2mn & 2mn & m^2 - n^2 \end{bmatrix} \begin{Bmatrix} \varepsilon_{xx} \\ \varepsilon_{yy} \\ \gamma_{xy} \end{Bmatrix} \tag{3.9}$$

3.6 Poisson's Ratio

As indicated in Fig. 3.1, unrestrained materials generally exhibit a lateral displacement, v, when subjected to axial strain ε_{xx}. Most materials exhibit a reduction in thickness when subjected to tensile normal strain. Some composite materials, and a few other materials, exhibit an increase in thickness (at least in some direction) when subjected to tensile strain. The negative ratio of the lateral strain ε_{yy} to axial strain ε_{xx} is defined as Poisson's ratio ν. Thus, for applied normal strain ε_{xx}:

$$\nu_{xy} = -\frac{\varepsilon_{yy}}{\varepsilon_{xx}} \tag{3.10}$$

$$\nu_{xz} = -\frac{\varepsilon_{zz}}{\varepsilon_{xx}} \tag{3.11}$$

The minus sign in the definition ensures that Poisson's ratio is positive for most materials. Subscripts on Poisson's ratio are used to clearly identify which Poisson's ratio is under consideration when the material is not isotropic. For an isotropic material, Poisson's ratio is independent of direction, i.e., $\nu_{xy} \equiv \nu_{xz} \equiv \nu$. There is a range of Poisson's ratio values for most materials; for glass 0.21–0.27, for metals 0.21–0.36, and for rubber 0.50 (see McClintock and Argon 1966). For laminated fibrous composite materials, Poisson's ratio exhibits a wide range of values including values greater than 1.0 and negative values (See Chap. 16).

3.7 Plane Strain

Plane strain in an x–y plane is defined as the condition that all out-of-plane strains are identically zero, i.e.,:

$$\varepsilon_{zz} \equiv \gamma_{zx} \equiv \gamma_{zy} \equiv 0 \tag{3.12}$$

3.8 Exercises

3.8.1 What is the % strain if $\varepsilon_{xx} = 0.001$?

3.8.2 What is the shear strain if the corresponding change in a right angle is 1°.?

3.8.3 Use Eq. (3.5) to show that Eq. (3.4) is correct.

3.8.4 Show that the strain tensor is symmetric.

3.8.5 Determine an expression for the maximum shear strain $\gamma_{x'y'}$ in terms of given strains ε_{xx}, ε_{yy} *and* γ_{xy}

Appendix: Solutions

3.8.1 What is the % strain if

$$\varepsilon_{xx} = 0.001$$

Solution

$$\varepsilon_{xx} = 0.001 * 100 = 0.1\,\%$$

3.8.2 What is the shear strain if the corresponding change in a right angle is 1°?

Solution

$$1^\circ * \frac{\pi}{180} = 0.0175\frac{in}{in} = 1.75\,\%$$

3.8.3 Use (3.5) to show that (3.4) is correct.

Solution

Example, for $i = 1, j = 2$, Eq. (3.5) $\Rightarrow$

$$\varepsilon_{12} = \frac{1}{2}\left(\frac{\partial u_2}{\partial x_1} + \frac{\partial u_1}{\partial x_2}\right) = \frac{1}{2}\gamma_{12}$$

3.8.4 Show that the strain tensor is symmetric, i.e., show that $\varepsilon_{12} = \varepsilon_{21}$.

Solution

$$\varepsilon_{12} = \frac{1}{2}\left(\frac{\partial u_2}{\partial x_1} + \frac{\partial u_1}{\partial x_2}\right)$$
$$\varepsilon_{21} = \frac{1}{2}\left(\frac{\partial u_1}{\partial x_2} + \frac{\partial u_2}{\partial x_1}\right)$$

3.8.5 Determine an expression for the maximum shear strain $\gamma_{x'y'}$ in terms of given strains ε_{xx}, ε_{yy} *and* γ_{xy}.
From Eq. (3.4)

$$\gamma_{x'y'} = -2(\varepsilon_{xx} - \varepsilon_{yy}) \sin\theta \cos\theta + \gamma_{xy}(\cos^2\theta - \sin^2\theta)$$

Solution
Setting the derivative with respect to θ equal 0 and using trigonometric identities:

$$\frac{d\gamma_{x'y'}}{d\theta} = 2(\varepsilon_{xx} - \varepsilon_{yy})\left[-\sin^2\theta + \cos^2\theta\right] + \gamma_{xy}[-4\sin\theta\cos\theta] = 0$$

$$\frac{d\gamma_{x'y'}}{d\theta} = (\varepsilon_{xx} - \varepsilon_{yy})\left[\frac{1+\cos 2\theta}{2} - \frac{1-\cos 2\theta}{2}\right] - 2\gamma_{xy}\sin\theta\cos\theta = 0$$

$$(\varepsilon_{xx} - \varepsilon_{yy})[\cos 2\theta] - \gamma_{xy}\sin 2\theta = 0$$

$$\tan 2\theta = \frac{(\varepsilon_{xx} - \varepsilon_{yy})}{\gamma_{xy}}$$

Reference

McClintock, F., & Argon, A. S. (1966). *Mechanical behavior of materials*. Addison-Wesley: Boston.

Chapter 4
Constitutive Equations

4.1 Normal Stress–Strain Response

Constitutive equations provide relationships between the stresses and the strains. For linear elastic materials, they are the so-called Hooke's law. In three-dimensional space, and for isotropic materials, the total normal strain in a given direction is the sum (superposition) of strains due to the three components of normal stress. Thus, they are written in terms of the stresses and two engineering constants, modulus, E, and Poisson's ratio, ν, as: Equation Section (Next)

$$\begin{aligned} \varepsilon_{xx} &= \frac{1}{E}\left(\sigma_{xx} - \nu\sigma_{yy} - \nu\sigma_{zz}\right) \\ \varepsilon_{yy} &= \frac{1}{E}\left(\sigma_{yy} - \nu\sigma_{xx} - \nu\sigma_{zz}\right) \\ \varepsilon_{zz} &= \frac{1}{E}\left(\sigma_{zz} - \nu\sigma_{xx} - \nu\sigma_{yy}\right) \end{aligned} \tag{4.1}$$

In the above, E is the *modulus of elasticity* (also called Young's modulus) and ν is Poisson's ratio described in the strain chapter. For a tensile test with the only nonzero applied stress σ_{xx}, the constitutive equation (4.1) reduces to $E = \frac{\sigma_{xx}}{\varepsilon_{xx}}$. Thus, the modulus of elasticity, E, is the slope of the stress versus strain diagram; it has the units of stress as strain is a dimensionless quantity.

Many materials, including most metals, exhibit nonlinear stress–strain behavior beyond the elastic range; nonlinear behavior is not depicted in Fig. 4.1. See Sect. 6. 5 for a brief discussion of nonlinear behavior. Within the linear elastic range, both loading and unloading exhibit linear elastic behavior. Thus, total unloading from a stress state in the linear range brings the state of stress and strain back to zero; there is no permanent deformation. Such is not the case if the loading extends beyond the linear range. As a result, linear elastic behavior is a much easier problem to solve.

C.T. Herakovich, *A Concise Introduction to Elastic Solids*,
DOI 10.1007/978-3-319-45602-7_4

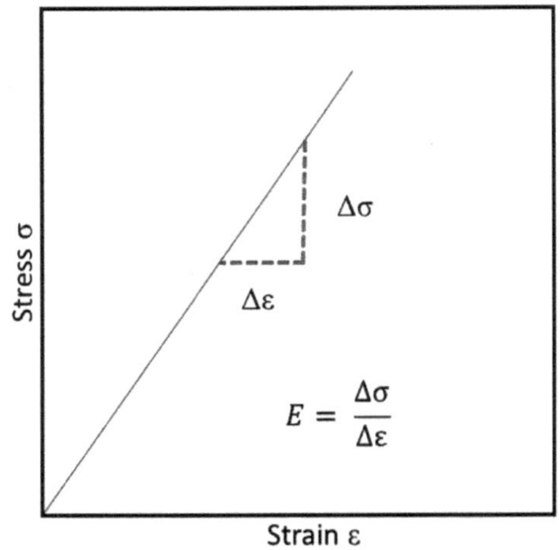

Fig. 4.1 Linear elastic stress–strain diagram

Most often, structures are designed to remain in the elastic range; exceptions are applications for crash worthiness and resistance to dynamic loadings such as explosions and earthquakes. In these cases, the desired outcome after loading is that the structure be intact, even if severely deformed.

4.2 Shear Stress–Strain Response

For shear response of an isotropic material, the (engineering) shear strains are written in terms of the *shear modulus*, G, (or equivalently in terms of E and ν) and the shear stresses as:

$$\begin{aligned} \gamma_{xy} &= \frac{1}{G}\tau_{xy} = \frac{2(1+\nu)}{E}\tau_{xy} \\ \gamma_{xz} &= \frac{1}{G}\tau_{xx} = \frac{2(1+\nu)}{E}\tau_{xy} \\ \gamma_{yz} &= \frac{1}{G}\tau_{yz} = \frac{2(1+\nu)}{E}\tau_{yz} \end{aligned} \tag{4.2}$$

The tensor shear strains are:

$$\varepsilon_{ij} = \frac{1}{2}\gamma_{ij} = \frac{1}{2G}\tau_{ij} = \frac{(1+\nu)}{E}\tau_{ij} \tag{4.3}$$

4.3 E, G, ν Relationship

From the above, it follows that:

$$G = \frac{E}{2(1+\nu)} \tag{4.4}$$

This relationship between E, G, and ν for isotropic materials can be arrived at through energy considerations (see, Boresi and Schmidt 2003, Sect. 3.3.2) or analysis of deformations in pure shear (see, Gere and Timoshenko 1997, Sect. 3.6). It is an important relationship because it shows that *there are only two independent constants for an isotropic, linear elastic material*. This means that the engineering properties (constants) for an isotropic, linear elastic material can be determined from one test in the laboratory. For a simple tension test with known applied axial (normal) stress, measurement of the axial strain gives the Young's modulus, E, and measurement of the lateral strain provides Poisson's ratio, ν.

The modulus, E, and Poisson's ratio, ν, for three commonly used metals are presented in Table 4.1 where the modulus values are given in terms of system international units, gigapascal (GPa), and US customary units, megastress (Msi) (note that 1.0 Msi = 6.8948 GPa). Clearly, steel is much stiffer than titanium and aluminum as demonstrated in Table 4.1.

4.4 Exercises

4.4.1 Verify the shear modulus values in Table 4.1 for the given E and ν.

4.4.2 Determine the normal strains ε_{xx} if $\sigma_{xx} = \sigma_{yy} = \sigma_{zz} = \sigma$.

4.4.3 If the strains are very small, their products may be neglected. What then is the change in volume of a cube subjected to the uniform normal stress σ as in Exercise 4.4.2?

4.4.4 For a plane state of stress in the x–y plane with known $\sigma_{xx} = \sigma_{yy} = \tau_{xy} = \sigma$, what is the z-components of normal strain ε_{zz}?

4.4.5 What are the shear strains γ_{xz} and γ_{yz} for the plane state of stress in 4.4.4?

Table 4.1 Modulus values for metals

Materials	Modulus E		Poisson's ratio	Shear modulus G	
	GPa	Msi	ν	GPa	Msi
Steel	200.0	29.0	0.32	132.0	19.1
Titanium	91.0	13.2	0.36	61.9	9.0
Aluminum	69.0	10.0	0.33	45.9	6.7

Appendix: Solutions

4.4.1 Verify the shear modulus values in Table 4.1 for the given E and v.

Solution

Using Excel and the relationship between E, G, and v gives:

Materials	E		v	G = E/2(1 + v)	
	Gpa	Msi		Gpa	Msi
Steel	200.0	29.0	0.32	132.0	19.1
Titanium	91.0	13.2	0.36	61.9	9.0
Aluminum	69.0	10.0	0.33	45.9	6.7

4.4.2 Determine the normal strains ε_{xx} if $\sigma_{xx} = \sigma_{yy} = \sigma_{zz} = \sigma$.

Solution

$$\varepsilon_{xx} = \frac{1}{E}\left(\sigma_{xx} - v\sigma_{yy} - v\sigma_{zz}\right)$$
$$\varepsilon_{xx} = \frac{\sigma}{E}(1 - 2v)$$

Note: if $v = \frac{1}{2}$, $\varepsilon_{xx} = 0$.

4.4.3 If the strains are very small, their products may be neglected. What then is the change in volume of a cube subjected to the uniform normal stress σ as in Exercise 4.4.2?

Solution

$$\varepsilon_{xx} = \frac{\sigma}{E}(1 - 2v) = \varepsilon_{yy} = \varepsilon_{zz} \equiv \varepsilon$$
$$\Delta V = (L+\varepsilon)^3 - L^3$$
$$\Delta V = L^3 + 3L^2\varepsilon + 3L\varepsilon^2 + \varepsilon^3 - L^3$$
$$\Delta V = 3L^2\varepsilon + 3L\varepsilon^2 + \varepsilon^3$$

Neglecting products of ε, ΔV, approaches $3L^2\varepsilon$.

4.4.4 For a plane state of stress in the x–y plane with known $\sigma_{xx} = \sigma_{yy} = \tau_{xy} = \sigma$, What is the Z-components of normal strain ε_{zz}?

Solution

$$\varepsilon_{zz} = \frac{1}{E}\left(\sigma_{zz} - v\sigma_{xx} - v\sigma_{yy}\right)$$
$$\varepsilon_{zz} = \frac{-2v\sigma}{E}$$

4.4.5 What are the shear strains γ_{xz} and γ_{yz} for the plane state of stress in 4.4.4?

Solution

$$\gamma_{xz} = \frac{1}{G}\tau_{xz} = \frac{2(1+v)}{E}\tau_{xz}$$
$$\gamma_{yz} = \frac{1}{G}\tau_{yz} = \frac{2(1+v)}{E}\tau_{yz}$$
$$\Rightarrow \gamma_{xz} = \gamma_{yz} = 0$$

References

Boresi, A. P., & Schmidt, R. J. (2003). *Advanced mechanics of materials* (6th ed.). New York: Wiley.

Gere, J. M., & Timoshenko, S. P. (1997). *Mechanics of materials* (4th ed.). Boston: PWS Publishing.

Chapter 5
Equilibrium

5.1 Newton's Second Law

Fundamental to the analysis of the response of solid materials is the application of Newton's second law (Newton 1687), namely that the summation of all external forces $\sum F$ is related to the mass, m, of the body and its acceleration, a, by:

$$\sum \mathrm{F} = ma \tag{5.1}$$

For bodies at rest (acceleration $a = 0$), this equation may be written in terms of the three components of force in the orthogonal x-, y-, and z-directions:

$$\begin{aligned} \sum F_x &= 0 \\ \sum F_y &= 0 \\ \sum F_z &= 0 \end{aligned} \tag{5.2}$$

Likewise, the equilibrium equations (zero angular acceleration) for moments about the three orthogonal axes are:

$$\begin{aligned} \sum M_x &= 0 \\ \sum M_y &= 0 \\ \sum M_z &= 0 \end{aligned} \tag{5.3}$$

Application of Newton's second law for equilibrium in terms of the stresses at a point (Fig. 2.1), with body forces assumed negligible in comparison with other forces, gives the results as partial differential equations (see Appendix):

C.T. Herakovich, *A Concise Introduction to Elastic Solids*,
DOI 10.1007/978-3-319-45602-7_5

$$\frac{\partial \sigma_x}{\partial x} + \frac{\partial \tau_{xy}}{\partial y} + \frac{\partial \tau_{xz}}{\partial z} = 0$$
$$\frac{\partial \sigma_y}{\partial y} + \frac{\partial \tau_{xy}}{\partial x} + \frac{\partial \tau_{yz}}{\partial z} = 0 \quad (5.4)$$
$$\frac{\partial \sigma_z}{\partial z} + \frac{\partial \tau_{xz}}{\partial x} + \frac{\partial \tau_{yz}}{\partial y} = 0$$

These equilibrium equations are fundamental and are used extensively for the analysis of bodies at rest.

5.2 Exercises

5.2.1 Determine the reactions at A and B for the simply supported beam loaded as shown.

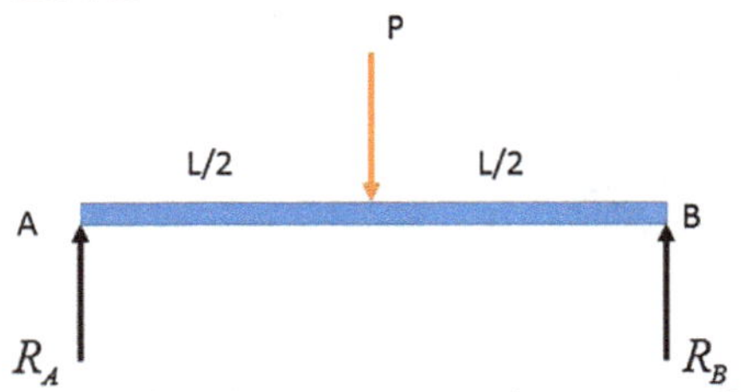

5.2.2 Determine the moment at the center of the beam in Exercise 5.2.1.

5.2.3 Determine the force reaction at A for the cantilevered beam shown below.

5.2.4 Determine the moment reaction at A for the cantilevered beam below.

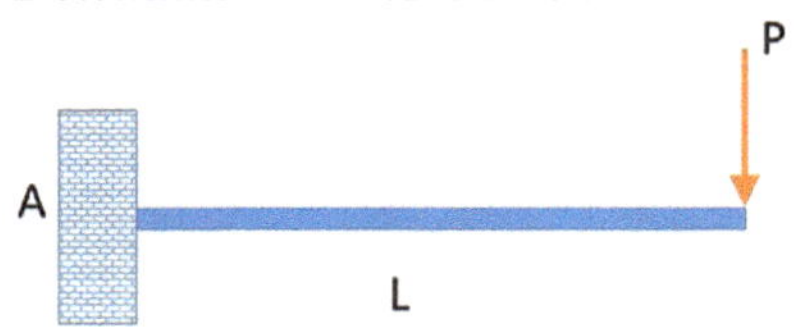

Appendix: Solutions

5.2.1 Determine the reactions at A and B for the simply supported beam loaded as shown.

Solution

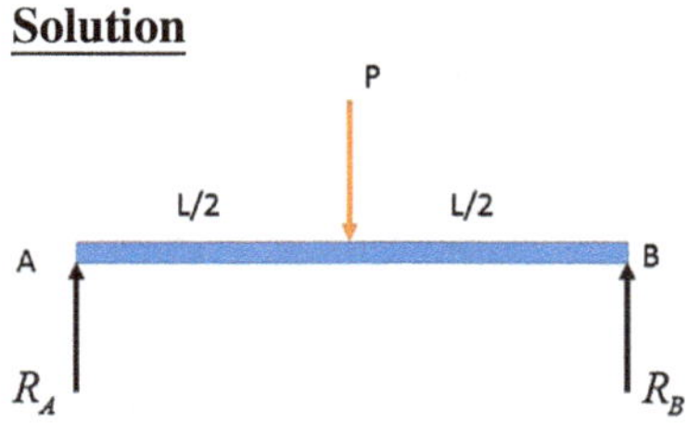

$$\sum M_A = 0 = P\frac{L}{2} - LR_B \quad \Rightarrow R_B = \frac{P}{2}$$
$$\sum F_y = 0 = R_A - P + R_B \quad \Rightarrow R_A = P - R_B$$
$$R_A = \frac{P}{2}$$

5.2.2 Determine the moment at the center of the beam in Exercise 5.2.1.

Solution

$$M_{x=\frac{L}{2}} = \frac{L}{2}R_A = \frac{L}{2}\frac{P}{2} = \frac{PL}{4}$$

5.2.3 Determine the force reaction at A for the cantilevered beam below.

Solution

$$\sum F_y = 0 \quad \Rightarrow R_A = P$$

5.2.4 Determine the moment reaction at A for the cantilevered beam below.

Solution

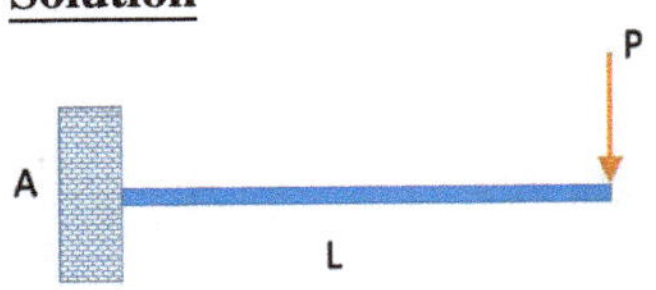

$$\sum M_A = 0 \quad \Rightarrow M_A = PL$$

Reference

Newton, I. (1729). *Mathematical principles of natural philosophy*, 1687 (A. Motte, Trans). Benjamin Motte.

Chapter 6
Axial Loading

6.1 Axial Stress

For axial stress loading, such as a tension test (see Fig. 3.1 repeated below as Fig. 6.1), of an isotropic, linear elastic material, the constitutive equation reduces to:

$$\sigma_{xx} = E\varepsilon_{xx} \tag{6.1}$$

where E is the elastic modulus of the material.

For an axial load, P, and a cross-sectional area, A, the axial normal stress, σ_{xx}, is:

$$\sigma_{xx} = \frac{P}{A} \tag{6.2}$$

6.2 Axial Strain

For a bar of original length, L, and a change in length, u, the axial strain, ε_{xx}, is:

$$\varepsilon_{xx} = \frac{u}{L} \tag{6.3}$$

6.3 Change in Length

Combining the equations for stress and strain (6.1–6.3) and integrating (adding up) the displacements over the length, L, give the total change in length, δ, of a rod subjected to axial loading:

C.T. Herakovich, *A Concise Introduction to Elastic Solids*,
DOI 10.1007/978-3-319-45602-7_6

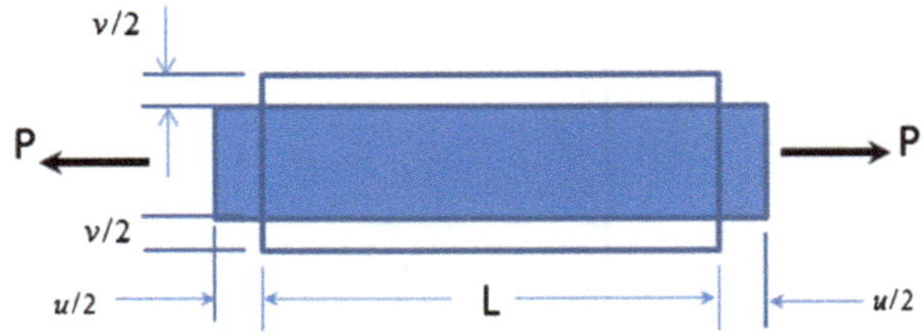

Fig. 6.1 Tensile loading of a bar

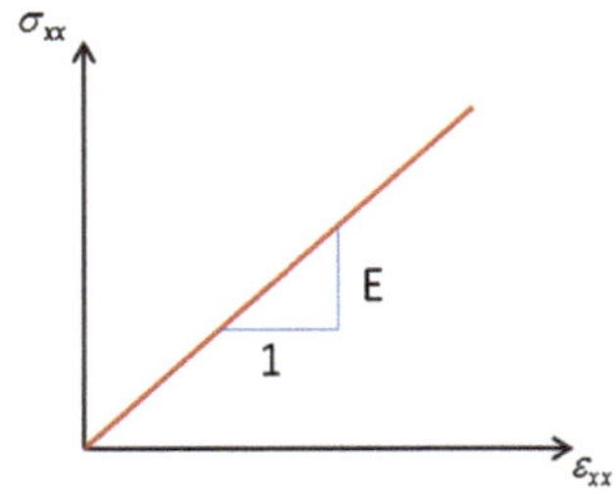

Fig. 6.2 Linear stress–strain response

$$\delta = \frac{PL}{AE} \tag{6.4}$$

This equation is fundamental for the design of a bar subjected to axial loading.

6.4 Stress–Strain Diagram

A plot of stress versus strain for the linear elastic range of a tension test is a straight line (Fig. 6.2). The slope of the stress–strain diagram is the elastic modulus E; modulus is also referred to as the stiffness of the material.

6.5 Nonlinear Response

Very few materials exhibit linear elastic stress–strain response up to failure. More typically, there is elastic response followed by nonlinear response as shown in Fig. 6.3. The demarcation between elastic and nonlinear response is defined as the *elastic limit* or *proportional limit*.

The modulus, E, and Poisson's ratio, ν, for three commonly used metals are presented in Table 4.1 (repeated here as Table 6.1) where the modulus values are given in terms of system international units, gigapascal (GPa), and US customary units, megastress (Msi) (note that 1.0 Msi = 6.8948 GPa). Clearly, steel is much stiffer than titanium and aluminum as demonstrated in Table 6.1 and Fig. 6.4.

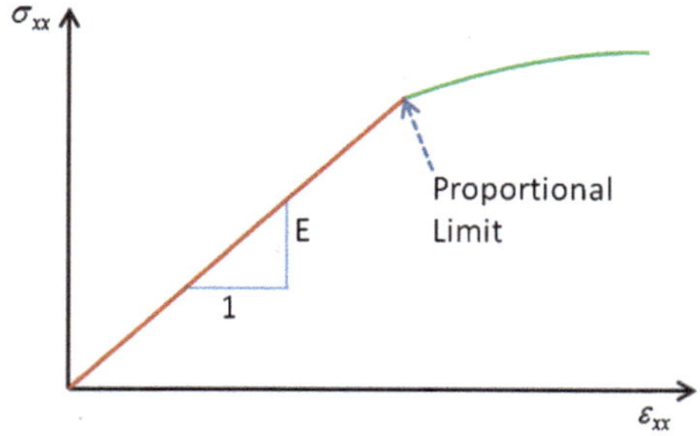

Fig. 6.3 Nonlinear stress–strain response

Table 6.1 Modulus values for metals

Material	Modulus E		Poisson's ratio	Shear modulus G	
	GPa	Msi	ν	GPa	Msi
Steel	200.0	29.0	0.32	132.0	19.1
Titanium	91.0	13.2	0.36	61.9	9.0
Aluminum	69.0	10.0	0.33	45.9	6.7

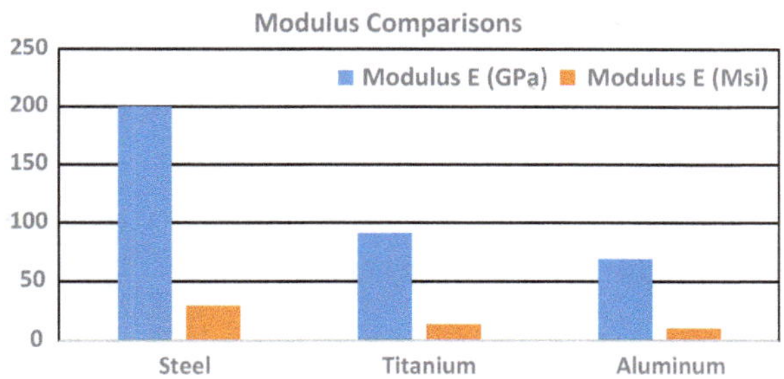

Fig. 6.4 Modulus comparisons

6.6 Exercises

6.6.1 A 0.5″-diameter steel rod is subjected to an axial load of 500 lbs. What is the stress in the rod?

6.6.2 If the rod in Exercise 6.6.1 is 5 ft. long, what is the change in length due to the 500-lb load?

6.6.3 What is the change in diameter of the rod in Exercise 6.6.2?

6.6.4 What is the stress in the rod in pascals?

6.6.5 What is the change in length in meters?

Appendix: Solutions

6.6.1 A 0.5″-diameter steel rod is subjected to an axial load of 500 lbs. What is the stress in the rod?

Solution

$$\sigma_{xx} = \frac{P}{A} = \frac{500}{\pi d^2/4} = \frac{2000}{\pi(0.5)^2} = 2{,}546.5\,\text{psi} = 2.546\,\text{ksi}$$

6.6.2 If the rod in Exercise 6.6.1 is 5 ft. long, what is the change in length due to the 500-lb load?

Solution

$$\delta = \frac{PL}{AE} = \frac{500(5 * 12)}{\pi(0.5)^2(29 * 10^6)} = 1317.1 * 10^{-6} = 0.0013171\ \text{in.}$$

6.6.3 What is the change in diameter of the rod in Exercise 6.6.2?

Solution

$$\Delta d = \varepsilon_{rr} d_1 = \nu \varepsilon_{xx} d_1 = (0.5)(0.32 * 0.0013171/60) = 3.5123 * 10^{-6}$$
$$\Delta d = 0.0000035123\,\text{in.}$$

6.6.4 What is the stress in the rod in pascals?

Solution

$$2,546.5\,\text{psi} * \frac{1\,\text{MPa}}{145\,\text{psi}} = 17.562\,\text{MPa} = 17.562 * 10^6\,\text{Pa}$$

6.6.5 What is the change in length in meters?

Solution

$$\delta\,\text{in.} * \frac{1.0\,\text{m}}{39.7\,\text{in.}} = \frac{0.0013171}{39.7} = 33.176 * 10^{-6}\,\text{m} = 33.176 * 10^{-4}\,\text{cm}$$
$$= 0.033176\,\text{mm}$$

Chapter 7
Torsion of Cylindrical Bars

Experimental evidence shows that when a solid, or hollow, cylindrical bar is subjected to equilibrated twisting moments (torque, T, applied at the opposite ends of the bar) (Fig. 7.1), the cross-sectional planes rotate through an angle of twist, θ, about the axis of the bar. The cross-sectional planes remain planar—they do not warp—and the radius of the bar does not change length. The angle of twist, θ, is proportional to the length, L, and the torque, T.

7.1 Shear Strain

The torque is the result of the shearing action applied at the ends of the bar. The shear strains associated with such torques can be determined by considering the displacements depicted in Fig. 7.2 where we consider a bar fixed at one end (or one-half of the bar shown in Fig. 7.1). For small deformations, the arc length B–B′ in the figure can be approximated from two triangles, one formed by the angle γ along the length of the bar and the other formed by the angle θ in the cross section. Thus,

$$\gamma = \frac{r\theta}{L} \tag{7.1}$$

The angle γ is the change in an original right angle and is the shear strain. From this equation, it is apparent that the shear strain at a given position, z = L, varies linearly with the radius, r, over the cross section of the bar.

C.T. Herakovich, *A Concise Introduction to Elastic Solids*,
DOI 10.1007/978-3-319-45602-7_7

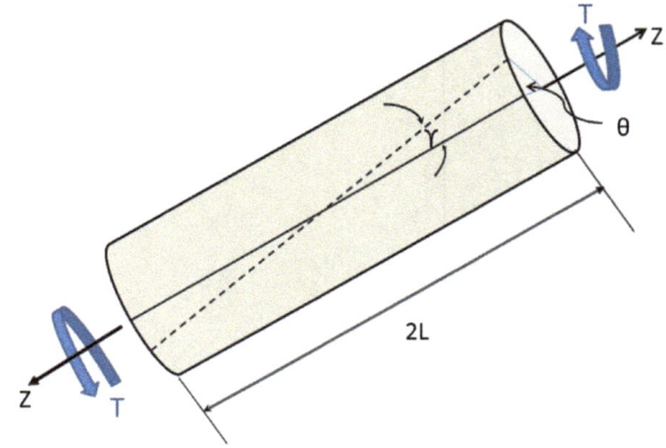

Fig. 7.1 Circular bar in torsion

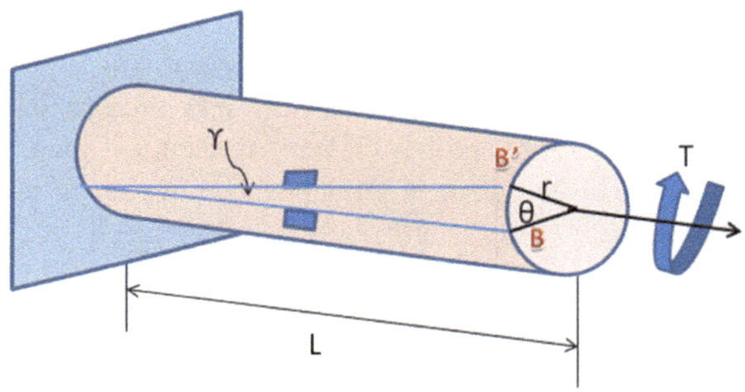

Fig. 7.2 Torsional displacements

7.2 Torque

When the bar is subjected to torque, shear stresses develop over the transverse cross-sectional area, A. The distribution of shear stresses (and shear strains) over the cross-sectional plane varies linearly with radial position, r, as shown in Fig. 7.3. The torque, T, about the z-axis, is the moment of the forces associated with the shear stresses $\tau_{z\theta}$, integrated over the cross-sectional area:

$$T = \int_A r\tau_{z\theta} dA \tag{7.2}$$

7.3 Angle of Twist

From the constitutive equation for pure shear loading and the two above equations, the torque, T, is:

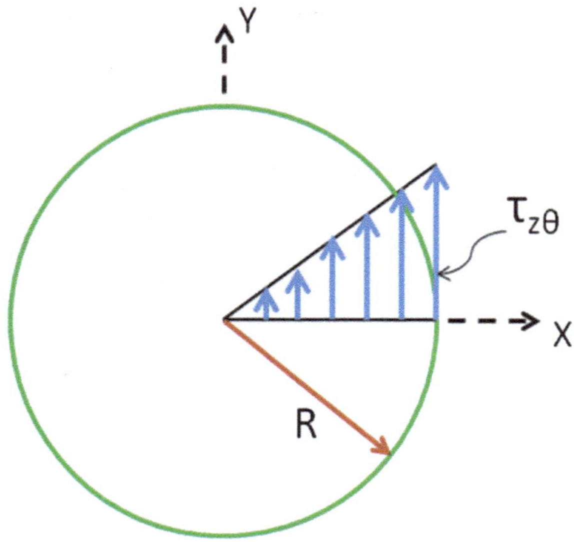

Fig. 7.3 Shear stress distribution in a circular bar

$$T = \frac{G\theta}{L}\int_A r^2 dA = \frac{G\theta J}{L} \tag{7.3}$$

where J is the polar moment of inertia of the cross-sectional area, A. Thus, for an applied torque T, the angle of twist is:

$$\theta = \frac{TL}{JG} \tag{7.4}$$

This equation is fundamental for design of circular bars subjected to torsion.

7.4 Maximum Shear Stress

The maximum shear stress in a circular cylinder under torsion is a fundamental design consideration. For a cylinder of radius, R, the maximum shear stress is determined by combining the constitutive equation, the strain equation, and the angle of twist equation with the result:

$$\tau_{z\theta}^{\max} = G\gamma^{\max} = \frac{GR\theta}{L} = \frac{TR}{J} \tag{7.5}$$

From the preceding development, it is evident that the analysis holds for both solid and hollow circular cylinders subjected to torsional loading.

7.5 Non-circular, Prismatic Bars

Methods are available for analyzing torsion of non-circular bars. The most general methods fall into the category of computational methods such as the finite element to be discussed in Chap. 15.

7.6 Exercises

7.6.1 Determine the polar moment of inertia of a circular bar radius R.

7.6.2 What torque is required to twist an aluminum bar through an angle of 10° if the bar is 2 cm in diameter and 1.5 m long?

7.6.3 What is the maximum shear stress in Pa for the bar in Exercise 7.6.2?

7.6.4 What is the midlength shear strain at a radius r = 0.5 cm for the bar in Exercise 7.6.2?

7.6.5 Compare the maximum stress in steel, titanium, and aluminum bars for the parameters of Exercise 7.6.2.

Appendix: Solutions

7.6.1 Determine the polar moment of inertia of a circular bar radius R

Solution

$$J = \int_0^{2\pi}\int_0^R r^2 r dr d\theta = 2\pi \int_0^R r^3 dr = \frac{2\pi R^4}{4} = \frac{\pi R^4}{2}$$

7.6.2 What torque is required to twist an aluminum bar through an angle of 10° if the bar is 2 cm in diameter and 1.5 m long?

Solution

$$
\begin{aligned}
\theta &= \frac{TL}{JG} \quad \Rightarrow \\
T &= \frac{\theta JG}{L} = \frac{10 * (\pi/180)(\pi R^4/2)\left(45.9 * 10^6\,\mathrm{N/m^2}\right)}{1.5\,m} \\
T &= \frac{10 * (\pi)\left(\pi(1\,\mathrm{cm} * 1\,\mathrm{m}/100\,\mathrm{cm})^4\right)\left(45.9 * 10^6\,\mathrm{N}\right)}{180 * 2 * 1.5\,\mathrm{m}^3} \\
T &= \frac{10 * (\pi)(\pi(1 * 1\,\mathrm{m}^4))\left(45.9 * 10^6\,\mathrm{N}\right)}{180 * 2 * 1.5\,m^3 * 100^4} = \frac{10 * \pi^2 * 45.9 * 10^6}{180 * 2 * 1.5 * 10^6}\,\mathrm{N-m} \\
T &= \frac{\pi^2 * 459.0}{180 * 3} = 2.097\,\mathrm{N-m}
\end{aligned}
$$

7.6.3 What is the maximum shear stress in Pa for the bar in Exercise 7.6.2?

Solution

$$
\tau^{\max} = \frac{TR}{J} = \frac{2.097 * R}{\pi R^4/2} = \frac{4.195}{\pi R^3} = \frac{4.195}{\pi\left(1 * 10^{-2}\right)^3} = 1.3352 * 10^6\,\mathrm{Pa} = 1.3352\,\mathrm{MPa}
$$

7.6.4 What is the midlength shear strain at a radius r = 0.5 cm for the bar in Exercise 7.6.2?

Solution

$$
\gamma = \frac{r\theta}{L} \quad \text{and} \quad \theta = \frac{TL}{JG} \Rightarrow \gamma = \frac{r}{L}\theta = \frac{r}{L}10 * \frac{\pi}{180} = \frac{0.5}{18 * 100 * 0.75} = 0.0004
$$

7.6.5 Compare the maximum stress in steel, titanium, and aluminum bars for the parameters of Exercise 7.6.2

Solution

$$
\tau^{\max} = \frac{TR}{J} \quad \text{and} \quad T = \frac{\theta JG}{L} \Rightarrow \tau^{\max} = \frac{\theta JG}{L}\frac{R}{J} = \frac{\theta R}{L}G
$$

Now, $\frac{\theta R}{L}$ is the same for all three bars. Thus,

$$\tau_{Al}^{\max} = \frac{\theta R}{L} G_{Al} \quad \tau_{St}^{\max} = \frac{\theta R}{L} G_{St} \quad \tau_{Ti}^{\max} = \frac{\theta R}{L} G_{Ti}$$

Chapter 8
Beam Bending

A beam is a structural member that is subjected to force and/or moment loading transverse to its length. The length of a beam is usually greater than both its width and its depth. An example of a beam under concentrated force loading is the diving board shown in Fig. 8.1. The diving board is a cantilevered beam subjected to a concentrated load (the diver) at its end. As shown in the figure, the beam bends under this loading.

8.1 Pure Bending

Pure bending of a beam refers to a beam, or a portion of a beam, where the loading corresponds to external moments only. There are no external normal or shear forces on this section of the beam. Figure 8.2 shows an example of pure bending in the section B–C of a simply supported beam subjected to concentrated force loading. From moment equilibrium $\sum M = 0$, we know that the bending moment diagram for this loading is shown in Fig. 8.2c. (The weight of the beam is neglected when it is small in comparison with other forces.)

8.2 Bending Moments

The distribution of stresses over the cross section of the beam in the section B–C must be equivalent to a moment about an axis that is perpendicular to the beam (the z-axis in Fig. 8.3). We consider a prismatic beam with a cross section that has a plane of symmetry (the x–y plane) with coordinate axes as defined in Fig. 8.3. Further, we take the x-axis to pass through the centroid of the cross section.

The magnitude of the moment of the normal stresses, σ_{xx}, about the z-axis must be equivalent to the external moment M. It is convention to define the moment such

C.T. Herakovich, *A Concise Introduction to Elastic Solids*,
DOI 10.1007/978-3-319-45602-7_8

Fig. 8.1 Diving board

Fig. 8.2 Pure bending in a beam subjected to concentrated loads

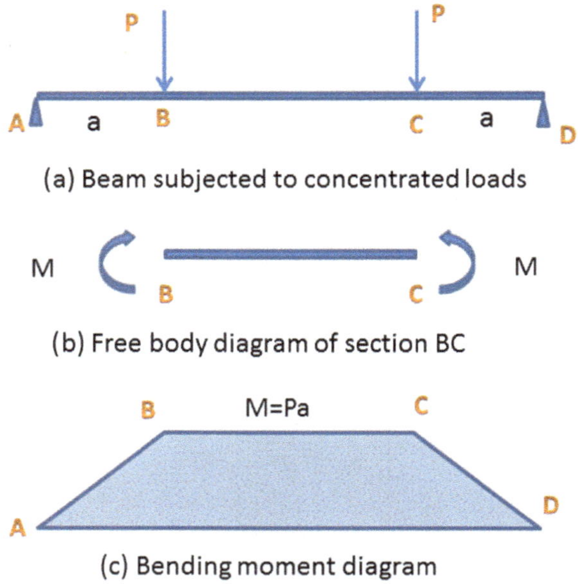

that tensile normal stress on the lower portion of the cross section results in a positive moment. Thus,

$$M = -\int_A y\sigma_{xx} dA \tag{8.1}$$

Fig. 8.3 Cross section of symmetric, prismatic beam with normal stress

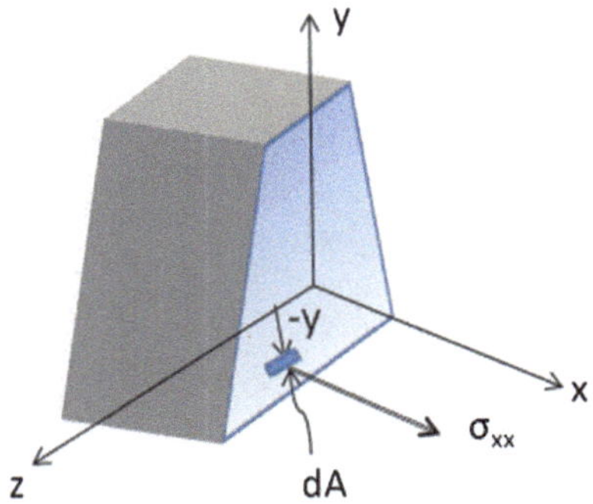

The distribution of the stresses over the cross section of the beam is related to the distribution of strains through Hooke's law. The strains are determined through the consideration of the curvature that develops in the beam as a result of the applied bending moment. The following section discusses the relationship between the strains and the beam curvature.

8.3 Beam Curvature and Strains

Figure 8.4 shows the deformation of a centroidal line passed through the plane of symmetry of a beam in pure bending. The originally straight lines along the length of the beam deform into circular arcs when subjected to equilibrated bending moments applied to the ends of the beam. Designating the radius of curvature of the line of zero strain (the neutral axis) as ρ, it can be shown (Gere and Timoshenko, Sect. 5.4) that the axial strain, ε_{xx}, at a position y above the neutral axis, is:

$$\varepsilon_{xx} = -\frac{y}{\rho} \tag{8.2}$$

Planes normal to the neutral axis prior to deformation remain plane and normal to the neutral axis after deformation as indicated in Fig. 8.5. It is evident from this figure that lines such as $a - a'$ above the neutral axis are compressed (negative strain) and lines such as $b - b'$ below the neutral axis are elongated (positive strain) for the positive bending moment shown. Thus, for the positive curvature shown in Figs. 8.4 and 8.5, strains above the neutral axis (positive y) are negative (compressive) and strains below the neutral axis (negative y) are positive (tensile).

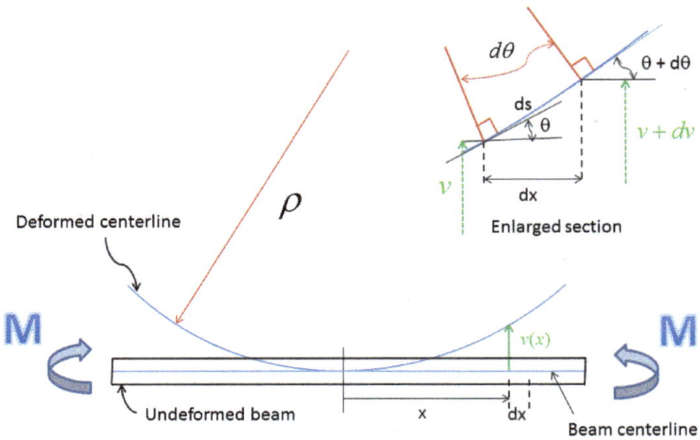

Fig. 8.4 Beam deflection

Fig. 8.5 Generic section of a beam in pure bending

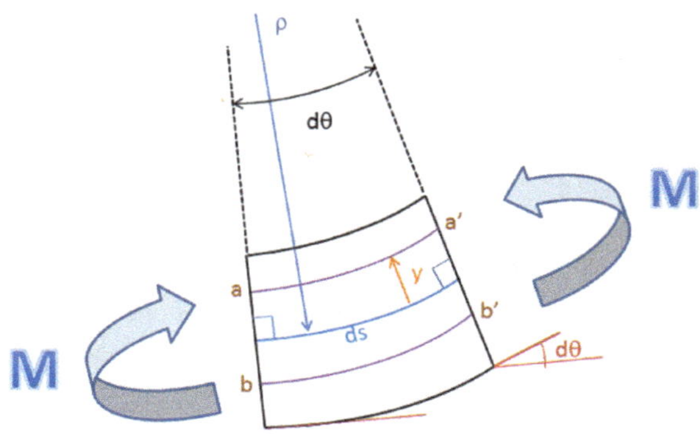

8.4 Stresses Due to Beam Bending

The stresses σ_{xx} associated with the axial strains are determined from Hooke's law (with all other components of stress being zero) as:

$$\sigma_{xx} = E\varepsilon_{xx} \tag{8.3}$$

Combining the above two equations gives the stresses as a function of the location, y, the material's elastic modulus, E, and the beam curvature, ρ, as:

$$\sigma_{xx} = -\frac{E}{\rho}y \tag{8.4}$$

We note here that it has been shown from the theory of elasticity, combined with experimental observations on the deformations in slender members subjected to small deformations, that the remaining five components of stress are zero for pure bending.

From the preceding equations, we see that the axial strain and stress are zero on the neutral axis (y = 0) and vary linearly with the distance y from the neutral axis. For a complete solution to the problem, it remains to determine the location of the neutral axis. The location can be determined as follows: For pure bending, the axis force is zero. The definition of axial force gives:

$$\int_A \sigma_{xx} dA = \int_A \left(-\frac{E}{\rho}\right) y dA = -\frac{E}{\rho}\int_A y dA = -\frac{E}{\rho}\bar{y}A = 0 \tag{8.5}$$

From the above, we see that the centroidal distance $\bar{y} = 0$ and the neutral axis passes through the centroid of the cross section.

For an explicit expression for the stresses in terms of the applied moment M, we consider the definition of moment:

$$M = -\int_A y\sigma_{xx} dA = \frac{E}{\rho}\int_A y^2 dA = \frac{E}{\rho} I \tag{8.6}$$

The minus sign is introduced in the above so that tensile stress below the axis gives rise to positive bending moment as per our definition. In the above equation, I is the moment of inertia of the cross section with respect to the z-axis.

From Eq. (8.6), we can write an expression for the curvature, $\rho(x)$, as a function of the bending moment, $M(x)$.

$$\rho(x) = \frac{EI}{M(x)} \tag{8.7}$$

Combining the stress and moment equations gives the stresses in terms of the applied moment, M, the moment of inertia, I, and the distance, y, from the neutral axis:

$$\sigma_{xx} = -\frac{M}{I} y \tag{8.8}$$

This equation is the fundamental equation for the axial stresses in a beam subjected to pure bending. It is evident that the maximum stress in a beam in pure bending occurs at the furthest point ($y = y^{\max}$) from the neutral axis. For positive bending moment, the maximum stress is compressive if it is above the neutral axis and it is tensile if it is below the neutral axis.

8.5 Exercises

8.5.1 Plot the variation of bending moment for the simply supported beam loaded as shown below.

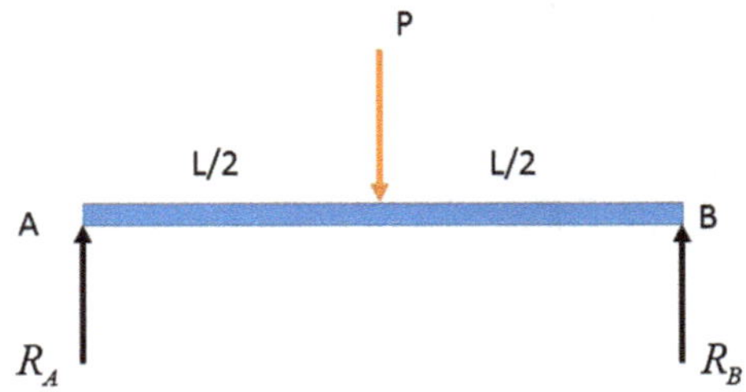

8.5.2 Plot the bending moment for the cantilevered beam loaded as shown below.

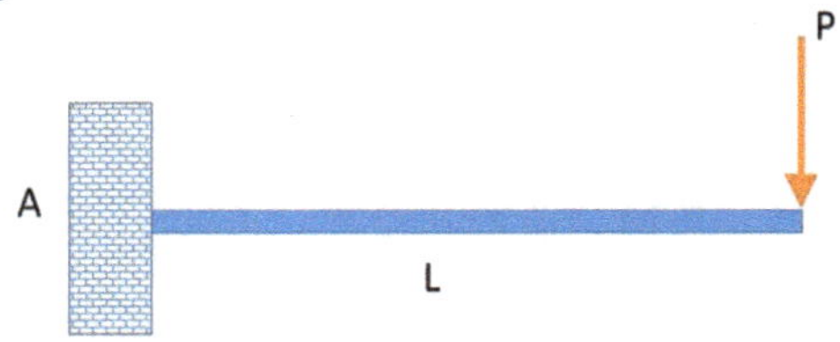

8.5.3 What is the curvature in an aluminum, 2″ square bar subjected to a bending moment M = 150 ft-lbs?

8.5.4 What is the maximum stress in the bar of Exercise 8.5.3?

8.5.5 What is the minimum stress in the bar of Exercise 8.5.3?

Appendix: Solutions

8.5.1 Plot the variation of bending moment for the simply supported beam loaded as shown.

<u>**Solution**</u>

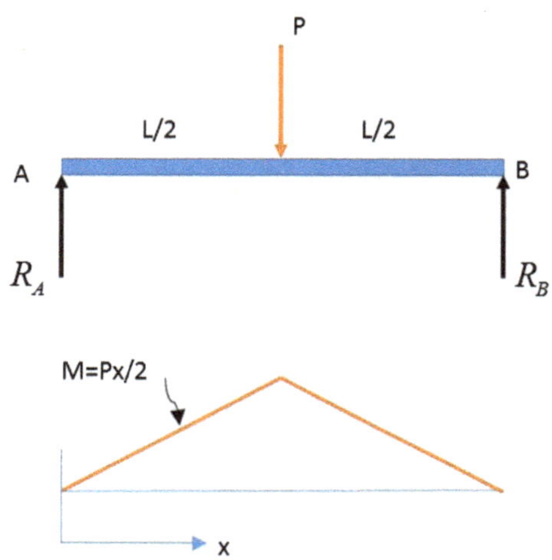

8.5.2 Plot the bending moment for the cantilevered beam loaded as shown.

Solution

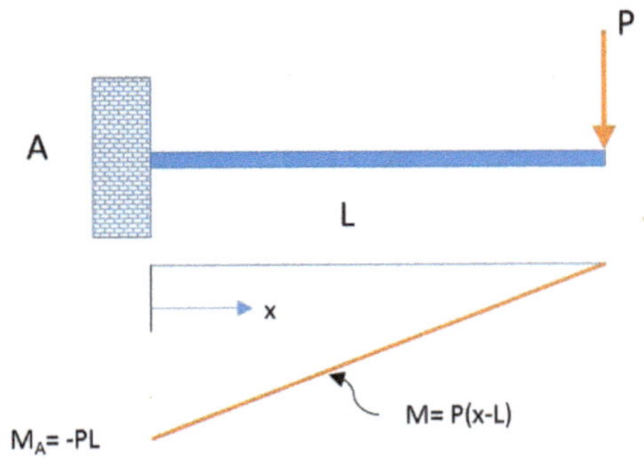

8.5.3 What is the curvature in an aluminum, 2″ square bar subjected to a bending moment M = 150 ft-lbs?

Solution
Recall Eq. (8.7):

$$\rho(x) = \frac{EI}{M(x)}$$

$$\rho = \frac{(10 * 10^6)\left(\frac{2*2^3}{12}\right)}{150 * 12} = \frac{(10 * 10^6)\left(\frac{16}{12}\right)}{150 * 12} = \frac{16 * 10^6}{15 * 12 * 12} = 0.0074 * 10^6$$
$$= 7{,}400 \text{ in.} = 616.67 \text{ ft.}$$

8.5.4 What is the maximum stress in the bar of Exercise 8.5.3?

Solution

$$\sigma_{xx} = -\frac{M}{I}y$$

For a positive moment, the maximum tensile stress is at the bottom of the beam:

$$\sigma_{xx}^{\max} = -\frac{150 * 12}{\frac{(2*2^3)}{12}}(-1) = \frac{150 * 144}{16} = 1{,}350 \text{ psi}$$

8.5.5 What is the minimum stress in the bar of Exercise 8.5.3?

Solution

The stress at the top of the bar is compressive (i.e., negative) and has the same magnitude as the stress at the bottom of the bar. Thus, $\sigma_{xx}^{Min} = -1,350\ psi$

Chapter 9
Beam Subjected to Transverse Loading

9.1 Equilibrium

Consider a statically determinate (all forces and moments can be determined directly from the equations of equilibrium) beam of prismatic cross section with a plane of symmetry passing through the centroid of the section (recall Fig. 8.3) and the lines of action of all external loads in this plane of symmetry. Figure 9.1 shows a simply supported beam subjected to a combination of transverse concentrated loads, P_i, and distributed loading, q(x).

Force and moment equilibrium equations can be applied to a free body diagram of the entire beam to determine the reactions at the supports. With the support reactions known, equilibrium of a free body diagram of a generic section of the beam, such as in Fig. 9.2, may be used to determine the transverse shear force, V (x), and bending moment, M(x), at any position x along the beam. We continue with the convention that positive moment gives tensile stresses below the neutral axis. For the shear force, V(x), convention is to define V as positive when the shear force is directed downward on a right-hand face as shown.

The variation of M(x) and V(x) along the length of the beam can be very helpful when assessing the entire beam for forces, moments, stresses, and deflections. As a first step, consider the forces and moments on an elemental beam section of length dx subjected to the distributed loading q(x) (Fig. 9.3).

Force equilibrium in the vertical direction gives the rate of change of the shear force as a function of the distributed load:

$$\frac{dV(x)}{dx} = -q(x) \tag{9.1}$$

Moment equilibrium of the elemental section gives the rate of change of moment as a function of the shear force:

C.T. Herakovich, *A Concise Introduction to Elastic Solids*,
DOI 10.1007/978-3-319-45602-7_9

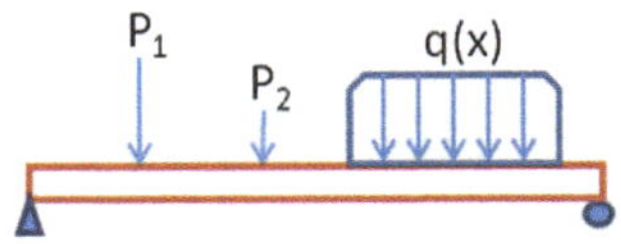

Fig. 9.1 Beam under transverse loads

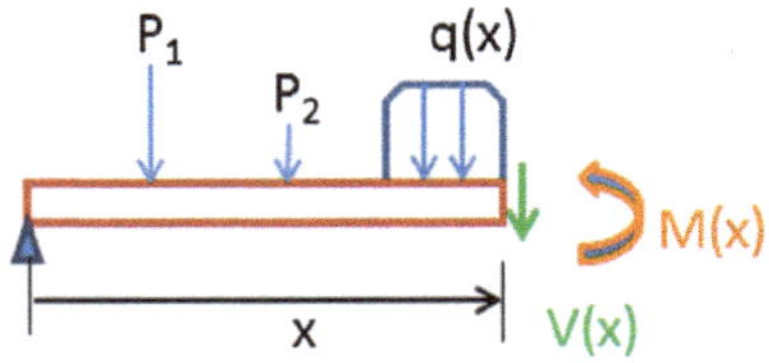

Fig. 9.2 Free body diagram of beam section

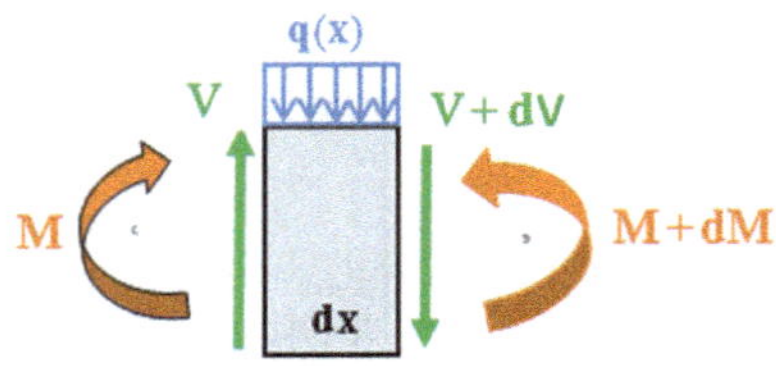

Fig. 9.3 Elemental beam section

$$\frac{dM(x)}{dx} = V(x) \tag{9.2}$$

These two equations must be modified in the cases of concentrated forces or concentrated moments acting at specific positions along the length of the beam (see Gere and Timoshenko, page 282 for details).

9.2 Stresses in Beams Under Transverse Loading

9.2.1 Normal Stress

The presence of moment M(x) and shear force V(x) at a generic cross section of a beam indicates that there must be normal stresses σ_{xx} and shear stresses τ_{xy} acting over the cross section (Fig. 9.4). The normal stresses are not affected by the presence of the shear stresses, and thus, they are defined in terms of the moment M (x) that is a function of position along the beam. From Eq. (8.7),

$$\sigma_{xx} = -\frac{M(x)}{I}y \tag{9.3}$$

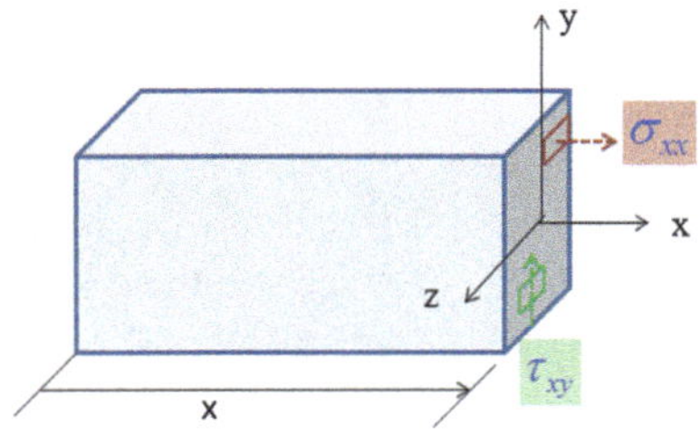

Fig. 9.4 Normal and shear stresses on a beam cross section

9.2.2 *Shear Stress*

The magnitude of the shear stresses must vary, as a function of the y location, over the transverse cross section. This is evident from the fact that the shear stress τ_{xy} must be nonzero over some portion of the cross section in order to provide the transverse shear force V; however, it must be zero on the top and bottom surfaces to satisfy the boundary condition of zero shear stress on these stress-free surfaces. Thus, the question is: How does τ_{xy} vary with y?

From the discussion of stress in Sect. 2.2, we know that if shear stresses are present on one face of a material element, they must also be present on other faces such that the stresses are equivalent to couples that render the element in equilibrium. Figure 2.2 shows the equilibrated shear stresses for plane stress. The answer to the above question is determined by considering a free body diagram of a transverse section of the beam of length dx and width b at $y = y_1$ (Fig. 9.5).

Equilibrium of forces in the axial (x) direction provides an expression for the shear stress at $y = y_1$.

$$\int_{A_1} \sigma_{xx} dA + \int_{A_2} \tau_{xy} b dx - \int_{A_3} \sigma_{xx} dA = 0 \tag{9.4}$$

In the above equation, A_1 and A_3 are the (equal) areas of the transverse sections above the line $y = y_1$ and A_2 is the area (ℓb) of the (horizontal) section at $y = y_1$, upon which the shear stress is acting. Introducing the equation for normal stress in terms of bending moment (9.3), the (algebraic) sum of integrals over A_1 and A_3 reduces to:

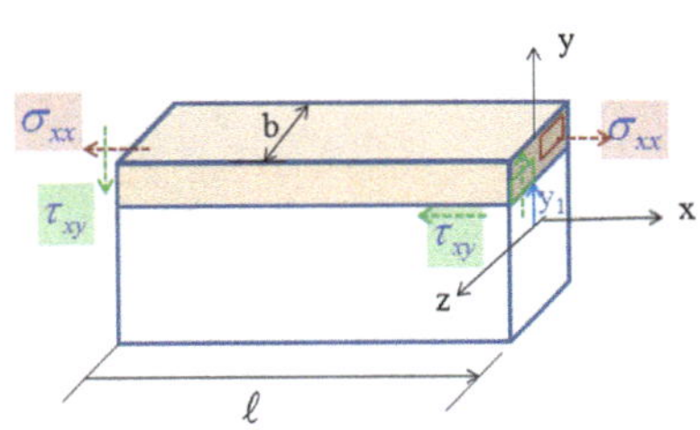

Fig. 9.5 Stresses in a section of a beam under transverse loading

$$\int_{A_1} -\frac{ydM}{I}dA \equiv -\frac{dM}{I}\int_{A_1} ydA \tag{9.5}$$

Now, the shear stress, τ_{xy}, is a specific value at $y = y_1$, and hence, combining (9.4) and (9.5) gives the shear stress as:

$$\tau_{xy} = \frac{1}{Ib}\left(\frac{dM}{dx}\right)\int_{A_1} ydA \tag{9.6}$$

Using $\frac{dM}{dx} = V(x)$ from (9.2), and defining the integral over the area A_1 above $y = y_1$ as $Q \equiv \int_{A_1} ydA$, the shear stress at location $y = y_1$ is given by:

$$\tau_{xy}\Big|_{y=y_1} = \frac{V(x)Q(y_1)}{Ib} \tag{9.7}$$

We note that this equation satisfies the boundary conditions $\tau_{xy} = 0$ on the top and bottom surfaces because $Q(y_1) = 0$ for these y values. This equation also indicates that the shear stress is a maximum along the beam at the x location corresponding to the maximum $V(x)$.

This Eq. (9.7) is fundamental for predicting the shear stress in a beam subjected to transverse loading.

9.3 Exercises

9.3.1 Plot the variation of shear force V along a beam loaded as shown below.

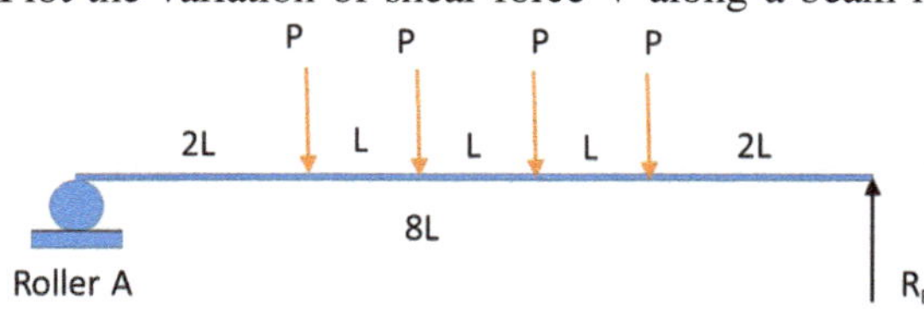

9.3.2 Plot the variation of moment M along the beam in Exercise 9.3.1.

9.3.3 Determine an expression for the maximum, tensile bending stress for a beam loaded as in 9.3.1 if the cross section of the beam is a rectangle, $2h$ height by $1h$ width.

9.3.4 Determine an expression for the maximum shear stress for the beam in Exercise 9.3.3.

9.3.5 What is the maximum compressive normal stress for the beam in Exercise 9.3.3, and where is it located?

Appendix: Solutions

9.3.1 Plot the variation of shear force V along a beam loaded as shown below.

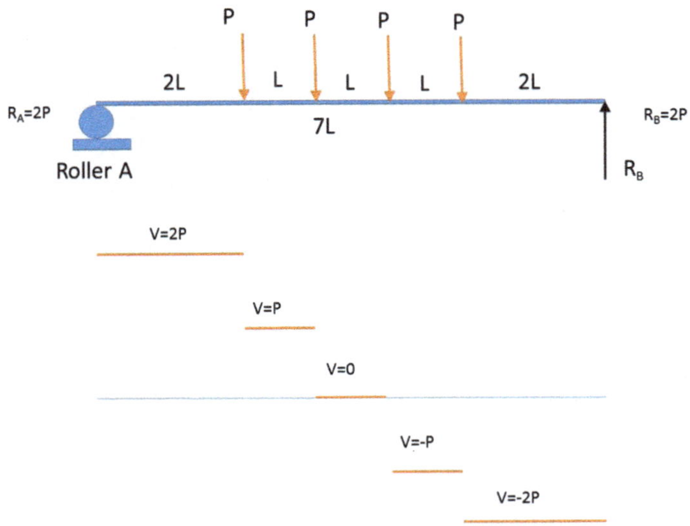

Solution

9.3.2 Plot the variation of moment M along the beam in Exercise 9.3.1.

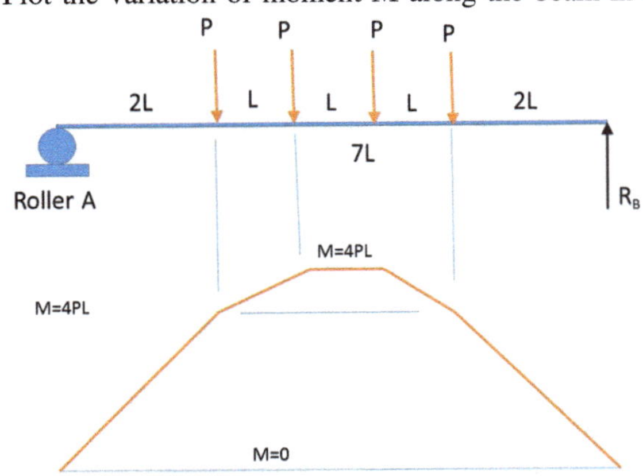

Solution

9.3.3 Determine an expression for the maximum, tensile bending stress for a beam loaded as in 9.3.1 if the cross section of the beam is a rectangle, $2h$ height by $1h$ width.

Solution

$$\sigma_{xx}^{\max} = \frac{-M^{\max} y^{\max}}{I} = \frac{-4PL(-h)}{\frac{1*h*h^3}{12}} = \frac{48PL}{h^3}$$

9.3.4 Determine an expression for the maximum shear stress for the beam in Exercise 9.3.3

Solution

$$\tau_{xy}\Big|_{y=0} = \frac{V(x)Q(y_1)}{Ib} = \frac{2PQ(y=o)}{\frac{1 * h * h^3 * h}{12}} = \frac{24P\left(\frac{h}{2} * h * h\right)}{h^5} = \frac{24P}{h^2}$$

9.3.5 What is the maximum compressive normal stress for the beam in Exercise 9.3.3 and where is it located?

Solution

Same magnitude as max tensile, but on top of the beam.

$$\sigma_{xx}^{\max} = \frac{-M^{\max} y^{\max}}{I} = \frac{-4PLh}{\frac{1 * h * h^3}{12}} = \frac{-48PL}{h^3}$$

Chapter 10
Beam Deflections

10.1 General Considerations

Beams that are subjected to loads deform with curvature as discussed in Chap. 8. The curvature is accompanied by displacements (deflections). For the coordinate system and loadings considered thus far, these deflections (designated, v), are in the vertical y-direction. Such deflections may be important for design. In the following we consider prismatic beams of isotropic, linear elastic materials under statically determinate bending moments and transverse loads such that the deflections are small and in the x–y plane of the beam. The x–y plane is a plane of symmetry. The problem is to determine $v(x)$, the variation of the vertical displacement, v, with location x along the beam. Since the beam is statically determinate, the support reactions can be determined from the static force and moments equilibrium equations.

Following Gere and Timoshenko (1997), the problem can be assessed by considering a cantilevered beam subjected to a concentrated load at the, otherwise, free end. Figure 10.1 shows the loaded beam and the deflection curve $v(x)$.

10.2 Deflection Equations

From Fig. 8.5, it is evident that the radius of curvature is related to an increment of arc length along the deflected curve by:

$$\rho d\theta = ds \tag{10.1}$$

In most situations, the beam deflections (v) and rotations (θ) are very small and we can use the approximation $ds \approx dx$. Thus:

C.T. Herakovich, *A Concise Introduction to Elastic Solids*,
DOI 10.1007/978-3-319-45602-7_10

Fig. 10.1 Cantilevered beam

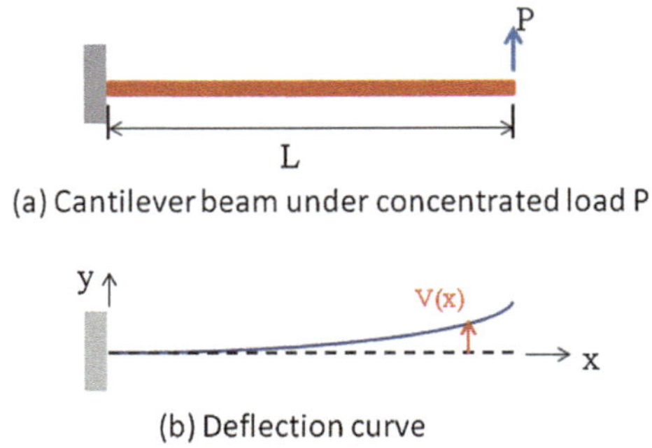

$$\frac{1}{\rho} = \frac{d\theta}{dx} \tag{10.2}$$

Further, for very small angles

$$\frac{dv}{dx} = \tan\theta \approx \theta \tag{10.3}$$

And thus, taking the derivative of θ in (10.3):

$$\frac{d\theta}{dx} = \frac{d^2v}{dx^2} \tag{10.4}$$

Combining (10.2 and 10.4) gives a relationship between the radius of curvature and the vertical deflection:

$$\frac{1}{\rho} = \frac{d^2v}{dx^2} \tag{10.5}$$

Combining (10.5) with the moment curvature relationship (8.6) gives the fundamental second-order differential relationship between deflection, v, moment, M, modulus, E, and moment of inertia, I.

$$EIv'' = M(x) \tag{10.6}$$

The primes (′) in (10.6) are used to denote differentiation with respect to x (e.g., $v'' = \frac{d^2v}{dx^2}$).

This derivative notation will be employed as we move forward to simplify the writing of equations.

Combining (10.5) and (10.6) gives the relationship between curvature and bending moment:

$$\frac{1}{\rho(x)} = \frac{M(x)}{EI} \tag{10.7}$$

Taking additional derivatives of (10.6) results in expressions in terms of the shear force, $V(x)$, and transverse loading, $q(x)$, where the results (9.1 and 9.2) have been employed:

$$EIv''' = V(x) \tag{10.8}$$

$$EIv'''' = -q(x) \tag{10.9}$$

This development assumes that all loading quantities are given function of the distance x along the beam. Integration of these differential equations introduces unknown constants which must be determined from known boundary conditions such as: deflection, v, slope, $dv/dx(v')$, shear force, V, or moment, M, at specified positions along the beam. Integrating the moment Eq. (10.6) gives:

$$EIv' = \int_x M(x)dx + C_1 \tag{10.10}$$

and a second integration gives:

$$EIv = \iint_x (M(x)dx)dx + \int_x C_1 dx + C_2 \tag{10.11}$$

where two constants of integration, C_1 and C_2 have been introduced.

From the two Eqs. (10.9 and 10.10), it is clear that M(x) must be a known function of x in order to perform the integrations. For statically determinate beams, M(x) can be determined from the equations of equilibrium. If the loading is not continuous over the length of the beam (such as introduction of concentrated loads at specific locations along the beam), the integration must be conducted over sections of the beam in which the loading is continuous. The complete solution for the beam deflection along the entire length of the beam is then obtained by matching slopes and deflections at the points of discontinuous loading.

Determination of the unknown constants follows directly from known boundary and continuity conditions. A simple support corresponds to $v = 0$, a fixed support corresponds to $v' = 0$; continuity conditions at points along the beam correspond to $v_1 = v_2$ and $v_1' = v_2'$ where the subscripts 1 and 2 correspond to locations on either side of a point, where the loading is discontinuous, such as where a concentrated load is applied.

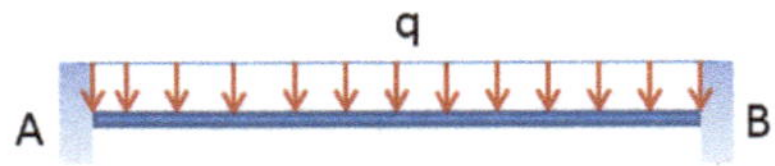

Fig. 10.2 Statically indeterminate beam

10.3 Statically Indeterminate Beams

It is often the case that the force and moment reactions at the supports cannot be determined directly from statics. Such beams are called *statically* indeterminate. The set of equations used to determine all unknowns must then include all boundary conditions and continuity conditions, in addition to the force and moment equilibrium equations. An example of a statically indeterminate beam is shown in Fig. 10.2, where a uniformly loaded beam is fixed at both ends.

There are six unknowns for such a loaded beam: force reactions at A and B, moment reactions at A and B, and two constants of integration C_1 and C_2. There are also six equations available to solve for the six unknowns: $v = 0$ at A and B, $v' = 0$ at A and B, overall force equilibrium $\sum F_y = 0$ $\sum F_y = 0$, and overall moment equilibrium $\sum M_z = 0$.

10.4 Exercises

10.4.1 Determine the bending moment, M, at A $(x = 0)$ for a cantilever beam, that is loaded as shown.

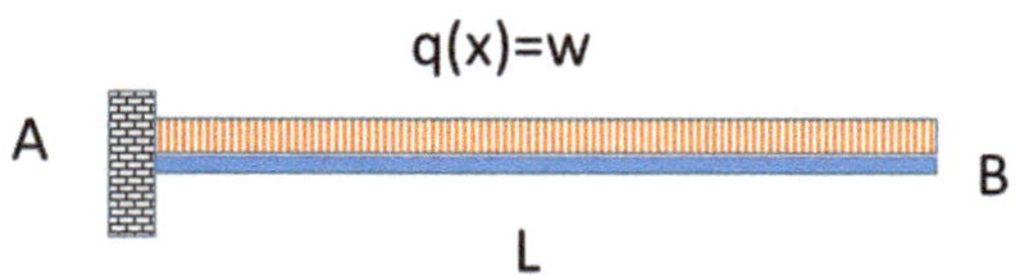

10.4.2 What are four boundary conditions for the beam in 10.4.1?

10.4.3 Determine the slope of the beam in exercise 10.4.1 at the end x = L.

10.4.4 Determine the displacement $v(L)$ at the end of the cantilevered beam loaded as shown above.

10.4.5 Determine the radius of curvature at $x = L$ for the beam in exercise 10.4.1.

Appendix: Solutions

10.4.1 Determine the bending moment, M, at A $(x = 0)$ for a cantilever beam loaded uniformly.

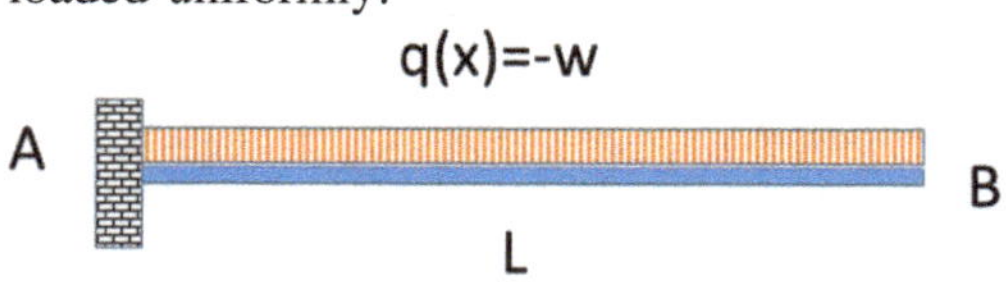

Solution

From moment equilibrium:

$$M_A = \int_0^L x(wdx) = \frac{-wL^2}{2}$$

10.4.2 What are four boundary conditions for the cantilever beam in 10.4.1?

Solution

$$v(0) = 0; \quad v'(0) = 0: \quad M(L) = 0; \quad V(L) = 0$$

10.4.3 Determine the slope of the cantilever beam in exercise 10.4.1 at the end x = L.

Solution

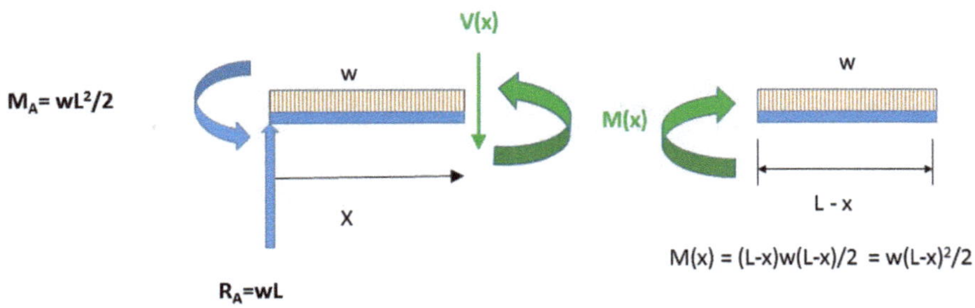

Free Body Diagrams of a portion of beam

From Eq. (10.10):

$$EIv'(x) = \int_0^x M(x)dx + C_1 = \int_0^x w\frac{(L-x)^2}{2}dx + C_1$$

$$EIv'(x) = \frac{w}{2}\frac{(L-x)^3}{3} + C_1 = \frac{w}{6}\left[L^3 - 3xL^2 + 3x^2L - x^3\right] + C_1$$

$$EIv'(0) = 0 = \frac{w}{6}L^3 + C_1 \quad \Rightarrow C_1 = -\frac{wL^3}{6}$$

$$EIv'(L) = \frac{w}{6}\left[L^3 - 3L^3 + 3L^3 - L^3\right] - \frac{wL^3}{6}$$

$$v'(L) = -\frac{wL^3}{6EI}$$

10.4.4 Determine the displacement $v(L)$ at the end of the cantilevered beam loaded as shown above.

Solution

Using the result from solution 10.4.3:

$$EIv'(x) = \frac{w}{2}\frac{(L-x)^3}{3} - \frac{wL^3}{6}$$

Integrating ⇒

$$EIv(x) = \frac{-w}{6}\frac{(L-x)^4}{4} - \frac{w}{6}L^3x + C_2$$

$$v(0) = 0 \quad \Rightarrow 0 = -\frac{w}{24}L^4 + C_2$$

$$\Rightarrow C_2 = \frac{wL^4}{24}$$

$$v(L) = -\frac{wL^4}{6EI} + \frac{wL^4}{24EI} = \frac{wL^4}{EI}\left(-\frac{4}{24} + \frac{1}{24}\right) = -\frac{3wL^4}{24EI}$$

$$v(L) = -\frac{wL^4}{8EI}$$

10.4.5 Determine the radius of curvature at $x = L$ for the beam in exercise 10.4.1.

Solution

Recall Eq. (10.5)

$$\frac{1}{\rho(x)} = \frac{M(x)}{EI}$$

At $x = L$, the moment M(L) is zero. Thus $\rho = \infty$, the beam is straight at the end x = L.

Reference

Gere, J. M., & Timoshenko, S. P. (1997). *Mechanics of materials* (4th ed.). PWS Publishing Co.

Chapter 11
Thermal Effects

11.1 Thermal Strains

Most materials experience extensional strains (change in size) when subjected to a temperature change. For unrestrained, isotropic materials, the extensional thermal strains, ε^T, are the same in all directions and can be expressed in terms of a coefficient of thermal expansion, α, and the temperature change, ΔT, as:

$$\varepsilon_{xx}^T = \varepsilon_{yy}^T = \varepsilon_{zz}^T = \alpha \cdot \Delta T \tag{11.1}$$

Isotropic materials do not exhibit shear strain when subjected to a temperature change. The thermal strains are often referred to as *free thermal strains.*

For a linear elastic material subjected to both mechanical strains (i.e., stress-related) and free thermal strains, the total strains, ε, are the sums of the mechanical strains, ε^σ, and the thermal strains, ε^T. Thus, the total x-direction strains are:

$$\varepsilon_{xx} = \varepsilon_{xx}^\sigma + \varepsilon_{xx}^T \tag{11.2}$$

11.2 Thermal Stresses

Substituting for the mechanical strains in terms of stresses (e.g., Eq. 6.1) and thermal strains in terms of $\alpha\Delta T$, we obtain the thermal stresses in terms of the total strains, ε_{xx}, the thermal strains, $\alpha\Delta T$, and the elastic modulus, E. For a one-dimensional state of stress σ_{xx}, the result is:

C.T. Herakovich, *A Concise Introduction to Elastic Solids*,
DOI 10.1007/978-3-319-45602-7_11

Fig. 11.1 Bar constrained between smooth walls

$$\sigma_{xx} = E(\varepsilon_{xx} - \alpha \Delta T) \tag{11.3}$$

As an example, consider a bar constrained between two smooth, frictionless walls (Fig. 11.1) and subjected to a temperature change ΔT, the total axial strain $\varepsilon_{xx} = 0$ and thus the axial stresses in the bar are compressive and given by:

$$\sigma_{xx} = -E\alpha \Delta T \tag{11.4}$$

It is assumed here that the bar is free to expand in the lateral directions because of the smooth, frictionless walls.

11.3 Exercises

11.3.1 Determine the change in volume of a cube, of length, h, that is subjected to a uniform temperature change ΔT if the cube is free to expand in all directions.

11.3.2 Determine the total change in length, δ, for an axial rod of length L that is subjected to a temperature change, ΔT, and an axial load, P, if the rod has modulus, E, coefficient of thermal expansion, α, Poisson's ratio, ν, and a cross-sectional area A.

11.3.3 What is the lateral strain ε_{yy} for the rod in Exercise 11.3.2?

11.3.4 What is the axial stress in a rod fixed at its ends between frictionless walls if it is subjected to a temperature change ΔT?

Appendix: Solutions

11.3.1 Determine the change in volume of a cube, of length, h, on a side that is subjected to a uniform temperature change ΔT if the cube is free to expand in all directions.

Solution

From Eq. (11.2)

$$\varepsilon_{xx} = \varepsilon_{xx}^{\sigma} + \varepsilon_{xx}^{T}$$

Likewise:

$$\varepsilon_{yy} = \varepsilon_{yy}^{\sigma} + \varepsilon_{yy}^{T}$$

$$\varepsilon_{zz} = \varepsilon_{zz}^{\sigma} + \varepsilon_{zz}^{T}$$

With all stresses being zero, the strains are the free thermal strains.

$$\varepsilon_{xx} = \varepsilon_{yy} = \varepsilon_{zz} = \alpha\Delta T$$

The expanded lengths are:

$$\Delta V = [h(1+\alpha\Delta T)]^3 - h^3$$
$$\Delta V = h^3(1+\alpha\Delta T)^3 - h^3$$
$$\Delta V = h^3\left(1+3\alpha\Delta T + 3\alpha^2\Delta T^2 + \alpha^3\Delta T^3\right) - h^3$$
$$\Delta V = h^3\left(3\alpha\Delta T + 3\alpha^2\Delta T^2 + \alpha^3\Delta T^3\right)$$

11.3.2 Determine the total change in length, δ, for an axial rod of length, L, that is subjected to a temperature change, ΔT, and an axial load, P, if the rod has modulus, E, coefficient of thermal expansion, α, Poisson's ratio, v, and a cross-sectional area A.

Solution

$$\delta = \frac{PL}{AE} + L\alpha\Delta T$$

11.3.3 What is the lateral strain ε_{yy} for the rod in Exercise 11.3.2?

Solution

$$\varepsilon_{yy} = \frac{1}{E}\left(\sigma_{yy} - v\sigma_{xx} - v\sigma_{zz}\right) + \alpha\Delta T$$

$$\varepsilon_{yy} = \frac{1}{E}\left(0 - v\frac{P}{A} - 0\right) + \alpha\Delta T$$

$$\varepsilon_{yy} = -\frac{vP}{AE} + \alpha\Delta T$$

11.3.4 What is the axial stress in a rod fixed at its ends between frictionless walls if it is subjected to a temperature change ΔT?

Solution

$$\varepsilon_{xx} = \frac{1}{E}\left(\sigma_{xx} - \nu\sigma_{yy} - \nu\sigma_{zz}\right) + \alpha\Delta T$$
$$\varepsilon_{xx} = \frac{1}{E}(\sigma_{xx} - 0 - 0) + \alpha\Delta T = 0$$
$$\sigma_{xx} = -E\alpha\Delta T$$

Chapter 12
Stability

Most large structures erected in earlier time, including the large cathedrals, temples, mosques, and synagogues, were erected as structures that were primarily under compression loading. When structural members are loaded in compression, there is a limiting value of the load for which the structure remains stable. A simple example of this structural response is a string; under tensile loading, the string is stable, but under compressive loading, the string is unstable, i.e., the string *buckles*.

As demonstrated in the following, the load at which a structural member buckles is a function of the material properties and geometry of the member. The load at which a structure buckles is called the *critical load*. If the structure is loaded below the critical load, it will return to stable equilibrium if displaced very slightly from the equilibrium position; however, if the structure is loaded slightly above the critical load, it will buckle when disturbed the smallest amount. The first successful analysis of a buckling problem was that by Euler (1744) for the case of a slender rod subjected to axial compressive load. In-depth studies of the stability of structures are given by Timoshenko and Gere (1961) and Bažant and Cedolin (1991).

12.1 Euler Buckling

Consider a long, slender column with pinned ends (i.e., free to rotate) under axial load, P, as shown in Fig. 12.1 (as in Gere and Timoshenko 1997). The load is applied through the centroid of the column cross section; the column is made of a linearly elastic material, is perfectly straight, of length, L, cross-sectional moment of inertia, I, and is assumed to have no imperfections. This problem is referred to as Euler buckling.

From the free body diagram in Fig. 12.1c, the bending moment Eq. (10.6) for the buckled column is:

C.T. Herakovich, *A Concise Introduction to Elastic Solids*,
DOI 10.1007/978-3-319-45602-7_12

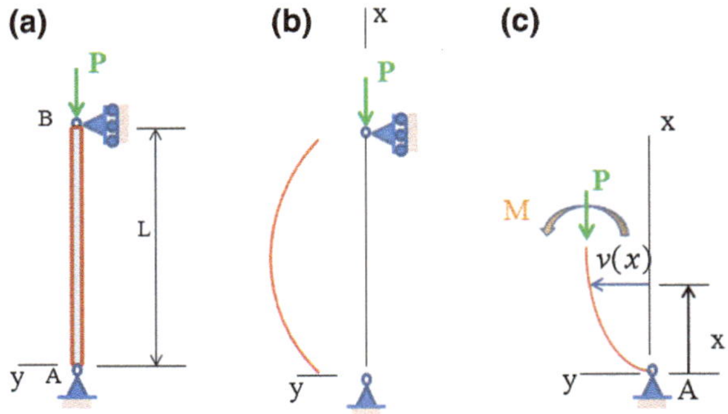

Fig. 12.1 Pinned-end column under axial compression. **a** Axially loaded pinned-end column. **b** Buckled column. **c** Internal reactions: buckled column

$$EIv'' = -Pv \tag{12.1}$$

It is convenient when solving (12.1) to combine terms and define:

$$k^2 = \frac{P}{EI} \tag{12.2}$$

Combining the above two equations gives the bending moment equation in the form:

$$v'' + k^2 v = 0 \tag{12.3}$$

The solution to this second-order differential equation is:

$$v = C_1 \sin kx + C_2 \cos kx \tag{12.4}$$

where C_1 and C_2 are constants of integration determined by the boundary conditions of the problem. For the pinned-end column of Fig. 12.1, the boundary conditions are as follows: The displacements are zero at the ends:

$$\begin{aligned} v(0) &= 0 \\ v(L) &= 0 \end{aligned} \tag{12.5}$$

From (12.4), the boundary condition at $x = 0$ requires that $C_2 = 0$. Thus, the shape of the deflection curve for the pinned-end rod is:

$$v(x) = C_1 \sin kx \tag{12.6}$$

The boundary condition at $x = L$ requires that $C_1 \sin kL = 0$. If C_1 and C_2 are both zero, the problem is trivial in that the deflection is zero. The non-trivial solution corresponds to sin $kL = 0$. Thus, for a non-trivial solution, it must be that:

$$kL = n\pi \quad n = 0, 1, 2, 3, \ldots \tag{12.7}$$

The case $n = 0$ corresponds to $v(x) \equiv 0$ and is of no interest as a buckling problem. Combining (12.2) and (12.7) gives the critical buckling load, P_{cr}, as:

$$P_{cr} = \frac{n^2\pi^2 EI}{L^2} \tag{12.8}$$

The smallest load P that satisfies the deflection equation is the case $n = 1$, and thus,

$$P_{cr} = \frac{\pi^2 EI}{L^2} \tag{12.9}$$

This equation is Euler's buckling equation for a pinned-end column. It shows that the critical load is inversely proportional to the length (L)-squared and proportional to the elastic modulus, E, and cross-sectional moment of inertia, I. Thus, long slender columns have low critical loads, and short, stout columns have high critical loads. The shape of the deflected column for the case n = 1 is a simple sine wave (Fig. 12.1b) with C_1 corresponding to the deflection at L/2. Higher values of n correspond to higher critical loads and higher buckling mode shapes.

The critical buckling load and the buckled shape of a column are influenced by the support conditions. Figure 12.2 shows a comparison of buckled columns for three types of supports, pinned–pinned, fixed–pinned, and fixed–fixed. Clearly, the fixed–fixed support condition results in the highest critical buckling load.

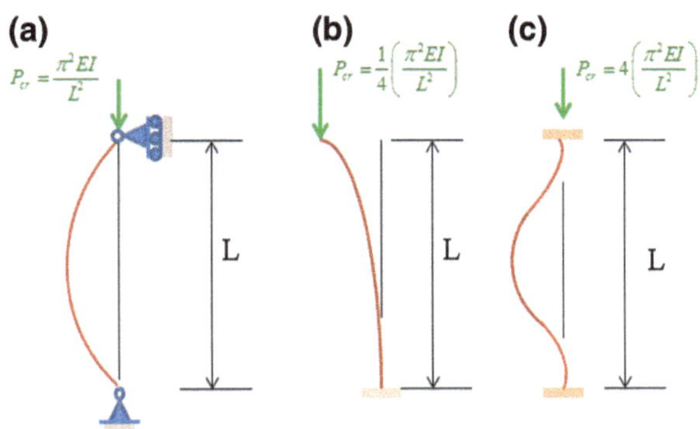

Fig. 12.2 Column support condition buckling modes. **a** Pinned–pinned. **b** Fixed–pinned. **c** Fixed–fixed

12.2 Exercises

12.2.1 Show that Eq. 12.4 is a solution to the second-order differential Eq. 12.3.

12.2.2 Compare the critical buckling loads for a pinned-end yardstick and a pinned-end, foot-long ruler of the same material and cross-sectional area.

Appendix: Solutions

12.2.1 Show that Eq. 12.4 is a solution to the second-order differential Eq. 12.3.

Solution

$$v'' + k^2 v = 0$$

$$v = C_1 \sin kx + C_2 \cos kx$$

$$\begin{aligned} v'(x) &= C_1 k \cos kx - C_2 k \sin kx \\ v''(x) &= -C_1 k^2 \sin kx - C_2 k^2 \cos kx = k^2(-C_1 \sin kx - C_2 \cos kx) \\ \Rightarrow v'' + k^2 v &= -k^2(C_1 \sin kx + C_2 \cos kx) + k^2(C_1 \sin kx + C_2 \cos kx) = 0 \end{aligned}$$

12.2.2 Compare the critical buckling loads for a pinned-end yardstick and a pinned-end, foot-long ruler of the same material and cross-sectional area.

Solution

$$P_{cr} = \frac{\pi^2 EI}{L^2}$$

$$\begin{aligned} P_{cr}^{YS} &= \frac{\pi^2 EI}{3^2} = \frac{\pi^2 EI}{9} \\ P_{cr}^{R} &= \frac{\pi^2 EI}{1^2} = \pi^2 EI \end{aligned}$$

The critical buckling load of the ruler is 9 times that of the yardstick.

References

Bažant, Z. P., & Cedolin, L. (1991). *Stability of structures.* Oxford: Oxford University Press.
Euler, L. (1933). *Methods inveniendi lineas curvas maximi minimive proprietate gaudentes …", Appendix I. "De curvis elasticis,* Bousquet, Lausanne and Geneva, 1744 (W. A. Oldfather, C. A. Ellis, & D. M. Brown, Trans), Isis, vol. XX.
Gere, J. M., & Timoshenko, S. P. (1997). *Mechanics of materials* (4th ed.). PWS Publishing Co.

Chapter 13
Thin-Walled Pressure Vessels

Many structures in use in everyday life are thin-walled vessels. As indicated in Fig. 13.1, thin-walled pressure vessels include a wide variety of configurations and uses. Examples include balloons and balls, hoses and pipes, small and large tanks, and portions of planes and space vehicles. A pressure vessel is considered thin-walled when the ratio of radius, r, to wall thickness, t, is 10.0 or greater. In the following, the stresses in thin-walled spheres and thin-walled cylinders subjected to uniform, internal pressure are studied.

13.1 Spherical Pressure Vessel

Consider a spherical thin-walled vessel made of linearly elastic material with modulus, E, and wall thickness, t. The inner radius of the sphere is r, and the sphere is under internal pressure, p. Since the wall thickness is small, we assume that the normal stresses throughout the wall thickness do not vary with position. Likewise, as the internal normal pressure is p, and the external normal pressure is zero, the through-the-thickness stresses are small in comparison with the normal stresses in the membrane.

The symmetry of the structure results in a uniform field of normal stresses acting at all points within the sphere walls. The shear stresses are zero throughout the sphere. For convenience, we designate this normal stress as σ_1 and refer to it as the *membrane stress*. Equilibrium of the free body diagram, in Fig. 13.2, then provides the fundamental relationship between the membrane stress, σ_1, the internal pressure, p, the sphere inner radius, r, and the wall thickness, t. It is noted that for thin spheres, the difference between the inner radius and the radius to the center of the wall is insignificant and can be ignored.

The summation of forces indicated in Fig. 13.2 equates the force associated with σ_1 acting around the circumference of the half sphere with the force associated with

C.T. Herakovich, *A Concise Introduction to Elastic Solids*,
DOI 10.1007/978-3-319-45602-7_13

Fig. 13.1 Thin-walled pressure vessels

Fig. 13.2 Free body diagram of half sphere

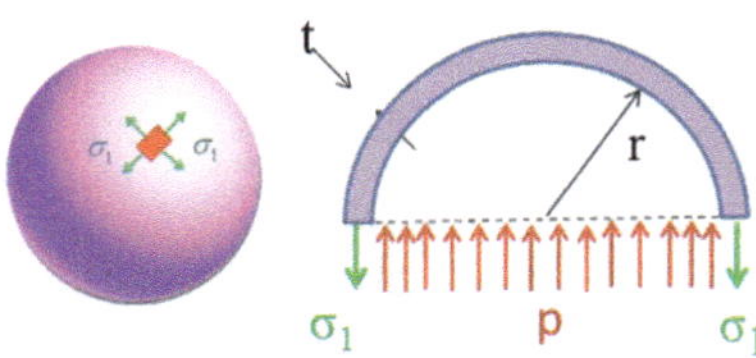

the pressure p acting over the flat circle of a slice through the center of the sphere. With the stresses known, the strains are determined from the constitutive equation.

$$\sigma_1 = \frac{pr}{2t} \tag{13.1}$$

13.2 Cylindrical Pressure Vessel

In the case of a cylindrical, thin-walled pressure vessel such as a pipe, we designate two normal stresses, σ_1, acting tangentially around the circumference of the cylinder and, σ_2, acting along the length of the cylinder (Fig. 13.3a). Equilibrium of the free body diagrams in Fig. 13.3 provides the relationships between the internal

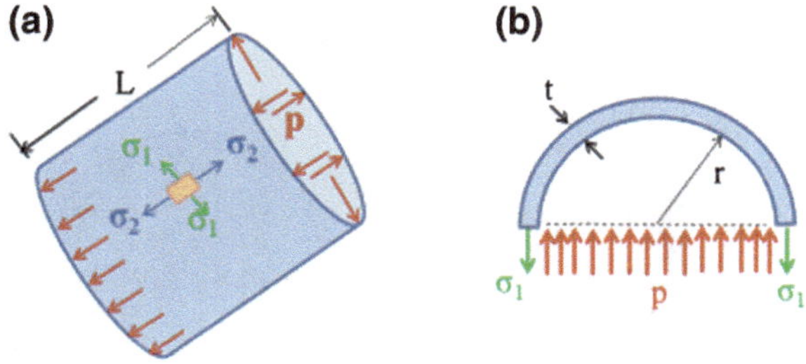

Fig. 13.3 Free body diagrams for a pressurized cylinder. **a** Pressurized. **b** Half cylinder: free body diagram

pressure, p, the axial stresses, σ_2, and the tangential stresses, σ_1. Equilibrium in the axial direction assumes that the ends of the cylinder are closed.

$$\sigma_1 = \frac{pr}{t} \tag{13.2}$$

$$\sigma_2 = \frac{pr}{2t} \tag{13.3}$$

Thus, for a cylinder under internal pressure, the tangential (or hoop) stress, σ_1, is double that of the axial stress, σ_2.

13.3 Exercises

13.3.1 Show that Eq. (13.1) is correct.

13.3.2 Determine the strains in a thin-walled sphere under internal pressure p in terms of the other sphere properties.

13.3.3 Show that Eq. (13.2) is correct.

13.3.4 Show that Eq. (13.3) is correct.

13.3.5 Basketballs used by men are generally larger in diameter than those used by women. If the balls are made of the same material and have the same internal pressure, what is the ratio of the thickness of a men's ball to that of the women's ball?

Appendix: Solutions

13.3.1 Show that Eq. (13.1) is correct.

Solution

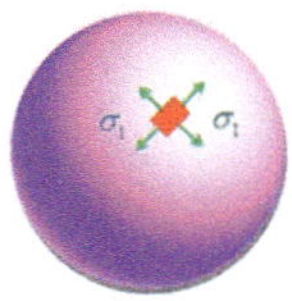

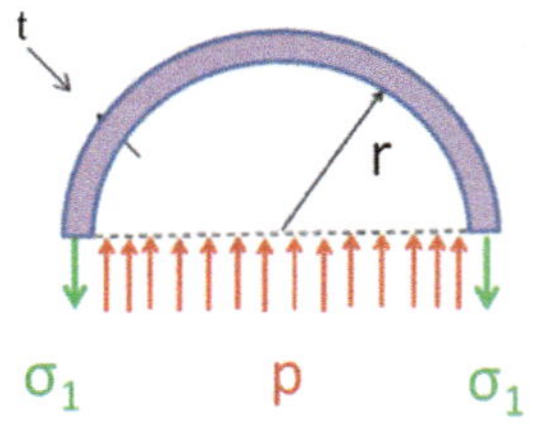

$$\sum F = 0$$
$$\sigma_1 2\pi rt = p\pi r^2$$
$$\sigma_1 = \frac{pr}{2t}$$

13.3.2 Determine the strains in a thin-walled sphere under internal pressure p in terms of the other sphere properties.

Solution

$$\sigma_1 = \sigma_2 = \frac{pr}{2t}$$
$$\varepsilon_1 = \frac{1}{E}(\sigma_1 - v\sigma_2) = \frac{1}{E}\frac{pr}{2t}(1 - v)$$
$$\varepsilon_2 = \varepsilon_1$$

13.3.3 Show that Eq. (13.2) is correct

Solution

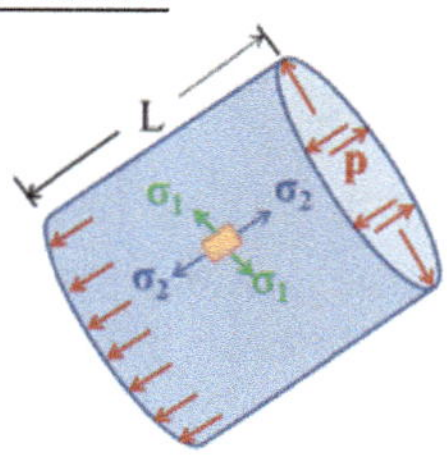

$$\sum F_2 = 0$$
$$\sigma_2 2\pi rt = p\pi r^2$$
$$\sigma_1 = \frac{pr}{2t}$$

a) Pressurized cylinder

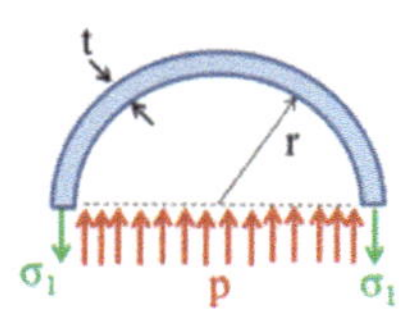

$$\sum F_1 = 0$$
$$\sigma_1 2Lt = p2rL$$
$$\sigma_1 = \frac{pr}{t}$$

b) Half cylinder: free body diagram

13.3.4 Show that Eq. (13.3) is correct

See the above-mentioned summation of forces.

13.3.5 Basketballs used by men are generally larger in diameter than those used by women. If the balls are made of the same material and have the same internal pressure, what is the ratio of the thickness of a men's ball to that of the women's ball if the stress is to be the same in both balls?

Solution

$$\sigma_1 = \sigma_2 = \frac{pr}{2t} \quad \Rightarrow \frac{p}{2\sigma} = \frac{t_M}{r_M} = \frac{t_W}{r_W} \quad \Rightarrow \frac{t_M}{t_W} = \frac{r_M}{r_W}$$

Chapter 14
Plates and Shells

14.1 Plates

A structural member that is in common usage is a flat plate. Floors, decks, manhole covers, and retaining walls are examples of flat plates. In general, a plate has a large plane area and a thickness that is much smaller than any dimension of the plane area. Plates can be subjected to a wide variety of loading and boundary conditions. Common examples of boundary conditions include simply supported, fixed, and free. Flat plates covering large areas are typically supported by equally spaced beams called "stringers." As a result of the variable loading and boundary conditions over the area of the plate, analysis of plates is more complicated than that of the idealized rods, beams, and thin-walled structural members analyzed previously in this treatise. An additional consideration for analysis of plates is the extent to which the deformations can be considered *small.* This is a valid assumption provided that the deformations are small in comparison with the plate thickness which is small in comparison with the lateral dimensions of the plate.

Analytical and approximate analytical solutions for a limited variety of plate geometries and loading conditions are discussed in Timoshenko and Woinowsky-Krieger (1959). For analysis of more complicated geometries, loadings, and support conditions, finite element methods or other approximate methods are used (see, e.g., Shames and Dym 1985).

A rectangular plate with associated bending moments, M, and shear forces, Q, (defined as moment and force intensities per unit length), is depicted in Fig. 14.1. Solutions to plate problems can be formulated in terms of these force and moment intensities, and the displacements, $w(x, y)$, of the plate midplane. We consider loading, $q(x, y)$, that is perpendicular to and variable over the surface of the plate. The moment M_{xy} is a twisting moment associated with the shear stresses in the plane of the plate. The shear forces Q_x and Q_y are associated with the transverse shear stresses τ_{xz} and τ_{yz}.

C.T. Herakovich, *A Concise Introduction to Elastic Solids*,
DOI 10.1007/978-3-319-45602-7_14

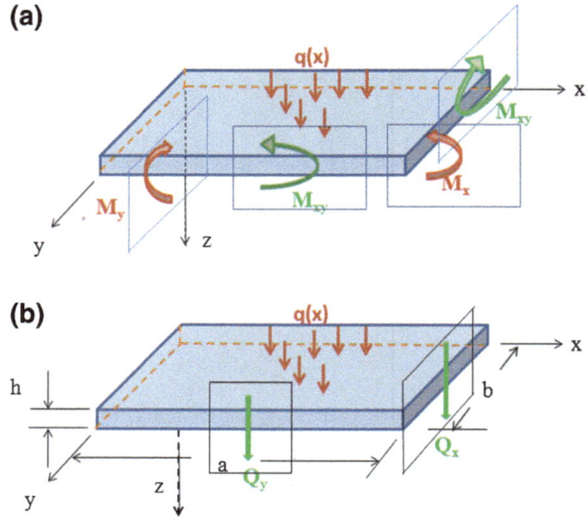

Fig. 14.1 Rectangular plate and intensities. **a** Plate bending moments per unit length. **b** Plate geometry and shear forces per unit length

The moment and shear intensities in Fig. 14.1 are defined in terms of stresses as follows:

$$
\begin{aligned}
M_x &= \int_{-h/2}^{h/2} \sigma_{xx} z dz \\
M_y &= \int_{-h/2}^{h/2} \sigma_{yy} z dz \\
M_{xy} &= \int_{-h/2}^{h/2} \tau_{xy} z dz \\
Q_x &= \int_{-h/2}^{h/2} \tau_{xz} dz \\
Q_y &= \int_{-h/2}^{h/2} \tau_{yz} dz
\end{aligned}
\tag{14.1}
$$

The vertical displacement of the plate midplane, $w(x, y)$, associated with bending, is depicted in Fig. 14.2 for the plate curvature in the x–z plane. A similar figure can be visualized for the y–z plane. It is assumed here, as was the case for bending

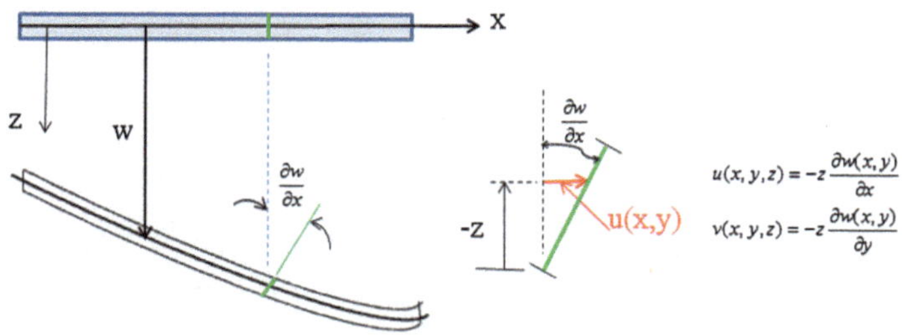

Fig. 14.2 Plate deformations associated with bending

of a beam, that lines perpendicular to the undeformed middle surface remain perpendicular to the deformed middle surface and do not change length. As indicated in the figure, the deformations $u(x, y, z)$ and $v(x, y, z)$ for the bending problem are given by:

$$\begin{aligned} u(x, y, z) &= -z\frac{\partial w(x, y)}{\partial x} \\ v(x, y, z) &= -z\frac{\partial w(x, y)}{\partial y} \end{aligned} \tag{14.2}$$

Using the above deformations (14.2) with the strain-displacement (3.2 and 3.3) and constitutive (4.1 and 4.2) equations for a homogeneous, linear elastic material, the fundamental equations describing the plate bending problem can be summarized as follows:

$$\begin{aligned} M_x &= -D\left(\frac{\partial^2 w}{\partial x^2} + v\frac{\partial^2 w}{\partial y^2}\right) \\ M_y &= -D\left(\frac{\partial^2 w}{\partial y^2} + v\frac{\partial^2 w}{\partial x^2}\right) \\ M_{xy} &= -(1 - v)D\left(\frac{\partial^2 w}{\partial x \partial y}\right) \end{aligned} \tag{14.3}$$

where D is the bending rigidity defined in terms of known parameters of the plate:

$$D = \frac{Eh^3}{12(1 - v^2)} \tag{14.4}$$

Equilibrium of an infinitesimal element such as that in Fig. 14.3 leads to the results:

$$\frac{\partial Q_x}{\partial x} + \frac{\partial Q_y}{\partial y} + q(x, y) = 0 \tag{14.5}$$

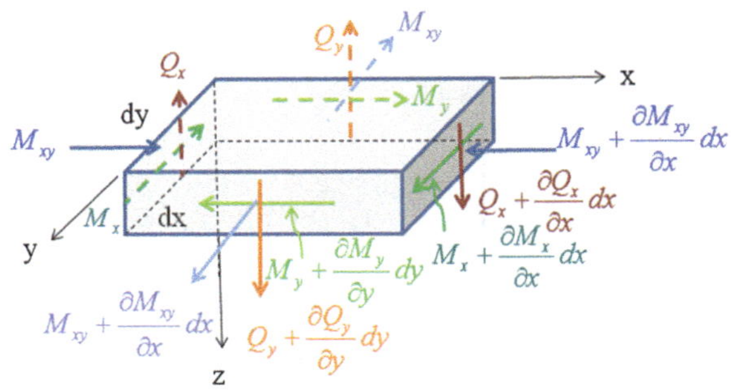

Fig. 14.3 Moments and shear forces on infinitesimal element

and

$$\frac{\partial^2 M_x}{\partial x^2}+2\frac{\partial^2 M_{xy}}{\partial x \partial y}+\frac{\partial^2 M_y}{\partial y^2}+q(x,y)=0 \tag{14.6}$$

Equation (14.6) can be expressed in terms of the vertical displacement $w(x,y)$ using (14.3) with the result (14.7). The operator ∇^4 is known as the *biharmonic operator* (see Appendix):

$$\nabla^4 w(x,y)\equiv\frac{\partial^4 w}{\partial x^4}+2\frac{\partial^2 w}{\partial x^2}\frac{\partial^2 w}{\partial y^2}+\frac{\partial^4 w}{\partial y^4}=\frac{q(x,y)}{D} \tag{14.7}$$

The above is the governing equation for classical plate theory. It was first presented in various incomplete forms by Sophie Germain during the period 1811–1816. Navier provided the complete form in an 1823 presentation to the French Academy. Finally, Kirchhoff published a paper (1850) where he clearly stated the assumptions used in developing the equations for small deformations of plates and beams. Specification of the boundary conditions and variation of load $q(x,y)$ must be known to solve the equation. Analytic solutions are available for only a limited number of cases.

14.2 Plate in Cylindrical Bending

One class of plate problems for which there is an analytical solution is a plate in cylindrical bending. Consider a plate (Fig. 14.4), that is long in the y-direction, simply supported along its edges and uniformly loaded $q(x,y)=q_{\circ}$ over the plate surface. The problem is then one-dimensional in x; the plate deforms in a cylindrical shape in the x–z plane. The boundary conditions for this problem are:

$$\begin{aligned} w(0)&=w(a)=0\\ M(0)&=M(a)=0 \end{aligned} \tag{14.8}$$

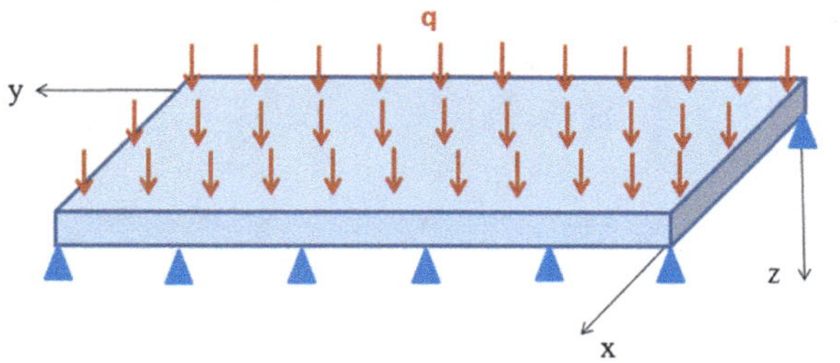

Fig. 14.4 Uniformly loaded long plate

These four boundary conditions provide the necessary equations for determination of the four unknown constants introduced upon integration of the fourth-order partial differential equation (14.7) (with partial derivatives with respect to y = 0). The general solution for displacement w(x) for the plate in cylindrical bending is:

$$w(x) = \frac{q_o x^4}{24D} + \frac{C_1 x^3}{6} + \frac{C_2 x^2}{2} + C_3 x + C_4 \tag{14.9}$$

The four boundary conditions result in the four constants:

$$\begin{aligned} C_1 &= -\frac{q_o a}{2D} \\ C_2 &= 0 \\ C_3 &= \frac{q_o a^3}{24D} \\ C_4 &= 0 \end{aligned} \tag{14.10}$$

Note that if the plate was fixed along its edges, the problem now would be solved using the boundary conditions:

$$\begin{aligned} &w(0) = w(a) = 0 \\ &\frac{\partial w(0)}{\partial x} = \frac{\partial w(a)}{\partial x} = 0 \end{aligned} \tag{14.11}$$

Representative deformation shapes for these two cases of boundary conditions are shown in Fig. 14.5.

14.3 Circular Plate Under Symmetrical Loading

A circular plate subjected to symmetrical loading, such as a distributed load $q(r)$ can be analyzed as a function of the radial coordinate r. Consider a thin, circular plate under loading $q(r)$ and uniformly supported at the edge $r = a$ (Fig. 14.6).

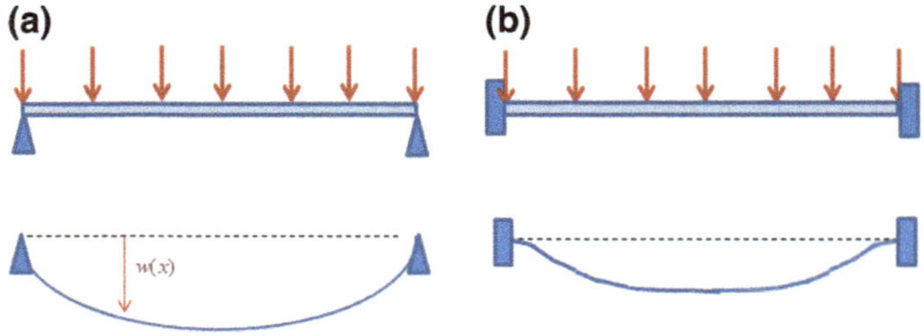

Fig. 14.5 Deformations in cylindrical bending. **a** Simply supported plate. **b** Fixed-end plate

Fig. 14.6 Circular plate

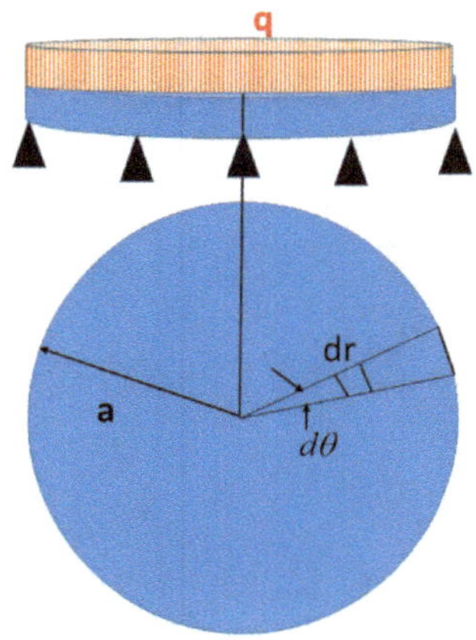

The differential equation for this problem (see Timoshenko and Woinowsky-Krieger 1959 for details) is:

$$\frac{1}{r}\frac{d}{dr}\{r\frac{d}{dr}[\frac{1}{r}\frac{d}{dr}(r\frac{dw}{dr})]\} = \frac{q(r)}{D} \tag{14.12}$$

For a uniform loaded $q(r) = q$, integration gives:

$$w(r) = \frac{qr^4}{64D} + C_1\frac{r^2}{4} + C_2 \log\frac{r}{a} + C_3 \tag{14.13}$$

The constants of integration $C_1, C_2 \, \text{and} \, C_3$ must be determined from the boundary conditions.

14.4 Circular Plate with Clamped Edge

The boundary conditions for this case are:

$$\frac{dw}{dr} = 0 \quad at\ r = 0\ \&\ r = a$$
$$w = 0 \quad at\ r = a \tag{14.14}$$

Solving for the unknown constants gives:

$$C_1 = -\frac{qa^2}{8D}$$
$$C_2 = 0 \tag{14.15}$$
$$C_3 = \frac{qa^4}{64D}$$

The final solution is then:

$$w(r) = \frac{q}{64D}\left(a^2 - r^2\right)^2 \tag{14.16}$$

14.5 Circular Plate with Simply Supported Edge

The boundary conditions for this case are:

$$\frac{dw}{dr} = 0 \quad at\ r = 0$$
$$w = 0 \quad at\ r = a \tag{14.17}$$
$$M_r(r = a) = 0$$

The final solution is then:

$$w(r) = \frac{q(a^2 - r^2)}{64D}\left[\left(\frac{5+v}{1+v}\right)(a^2 - r^2)\right] \tag{14.18}$$

14.6 Shells

Shells may be thought of as curved plates with the thickness being much smaller than the dimensions of the surface. Shell structures include naturally occurring egg shells and man-made canoes (Fig. 14.7). One example of a rather exotic shell structure is the Sydney Opera House (Fig. 14.8). It was designated a UNESCO World Heritage site in 2007. Basic elements of shell theory are presented in Timoshenko and Woinowsky-Krieger (1959). As is the case for plates, analytical solutions exist for a very limited number of idealized cases including the

Egg Shell Canoe

Fig. 14.7 Shell structures

Fig. 14.8 Sydney Opera House

thin-walled sphere and cylinder discussed in Chap. 13. Finite element and other approximate computational mechanics methods typically are used to analyze more complex shell structures.

In view of the additional complexity of shell structures, no attempt will be made here to analyze them more than was done in Chap. 13. For advanced treatment of shells, the reader is referred to the book by Libai and Simmonds (1998).

14.7 Exercises

14.7.1 Show that Eq. (14.7) is correct.

14.7.2 Show that Eq. (14.10) gives the unknown constants for a uniformly loaded plate in cylindrical bending.

14.7.3 Show that Eq. (14.13) is a solution of Eq. (14.12).
14.7.4 What is the maximum vertical deflection of the uniformly loaded plate in Fig. 14.4?

Appendix: Solutions

14.7.1 Show that Eq. (14.7) is correct.

$$\nabla^4 w(x,y) \equiv \frac{\partial^4 w}{\partial x^4} + 2\frac{\partial^2 w}{\partial x^2}\frac{\partial^2 w}{\partial y^2} + \frac{\partial^4 w}{\partial y^4} = \frac{q(x,y)}{D}$$

Solution

Substituting Eq. (14.3):

$$M_x = -D\left(\frac{\partial^2 w}{\partial x^2} + v\frac{\partial^2 w}{\partial y^2}\right)$$
$$M_y = -D\left(\frac{\partial^2 w}{\partial y^2} + v\frac{\partial^2 w}{\partial x^2}\right)$$
$$M_{xy} = -(1-v)D\left(\frac{\partial^2 w}{\partial x \partial y}\right)$$

into (14.6)

$$\frac{\partial^2 M_x}{\partial x^2} + 2\frac{\partial^2 M_{xy}}{\partial x \partial y} + \frac{\partial^2 M_y}{\partial y^2} + q(x,y) = 0$$

gives:

$$\frac{\partial^2\left(\frac{\partial^2 w}{\partial x^2} + v\frac{\partial^2 w}{\partial y^2}\right)}{\partial x^2} + 2\frac{\partial^2(1-v)\left(\frac{\partial^2 w}{\partial x \partial y}\right)}{\partial x \partial y} + \frac{\partial^2\left(\frac{\partial^2 w}{\partial y^2} + v\frac{\partial^2 w}{\partial x^2}\right)}{\partial y^2} = \frac{q(x,y)}{D}$$
$$\frac{\partial^4 w}{\partial x^4} + v\frac{\partial^4 w}{\partial x^2 \partial y^2} + 2(1-v)\frac{\partial^4 w}{\partial x^2 \partial y^2} + \frac{\partial^4 w}{\partial y^4} + v\frac{\partial^4 w}{\partial x^2 \partial y^2} = \frac{q(x,y)}{D}$$
$$\frac{\partial^4 w}{\partial x^4} + 2\frac{\partial^4 w}{\partial x^2 \partial y^2} + \frac{\partial^4 w}{\partial y^4} = \frac{q(x,y)}{D} \equiv \nabla^4 w$$

14.7.2 Show that Eq. (14.10) gives the unknown constants for a uniformly loaded plate in cylindrical bending.

$$
\begin{aligned}
C_1 &= -\frac{q_\circ a}{2D} \\
C_2 &= 0 \\
C_3 &= \frac{q_\circ a^3}{24D} \\
C_4 &= 0
\end{aligned}
$$

Solution

From Eq. (14.9), the general solution is:

$$w(x) = \frac{q_\circ x^4}{24D} + \frac{C_1 x^3}{6} + \frac{C_2 x^2}{2} + C_3 x + C_4$$

From Eq. (14.8), the boundary conditions are:

$$
\begin{aligned}
w(0) = w(a) = 0 \\
M_x(0) = M_x(a) = 0
\end{aligned}
$$

Substitutions give:

$$
\begin{aligned}
w(0) = 0 \quad & \Rightarrow C_4 = 0 \\
M_x(0) = 0 \quad & \Rightarrow -D\left(\frac{\partial^2 w}{\partial x^2}\right) = 0 \quad \Rightarrow C_2 = 0 \\
M_x(a) = 0 \quad & \Rightarrow \frac{(q_\circ a^2)}{2D} + C_1 a = 0 \quad \Rightarrow C_1 = -\frac{q_\circ a}{2D} \\
w(a) = 0 \quad & \Rightarrow \frac{q_\circ a^4}{24D} + \left(-\frac{q_\circ a}{2D}\right)\frac{a^3}{6} + C_3 a = 0 \quad \Rightarrow C_3 = \frac{q_\circ a^3}{24D}
\end{aligned}
$$

14.7.3 Show that Eq. (14.13) is a solution of Eq. (14.12) for a circular plate under uniform loading.

Recall Eq. (14.12):

$$\frac{1}{r}\frac{d}{dr}\{r\frac{d}{dr}[\frac{1}{r}\frac{d}{dr}(r\frac{dw}{dr})]\} = \frac{q(r)}{D}$$

and (14.13):

$$w(r) = \frac{qr^4}{64D} + C_1\frac{r^2}{4} + C_2 \log\frac{r}{a} + C_3$$

Solution

Substitution using the known values of the constants $C_2 = C_4 = 0$; $C_1 = \frac{-q_\circ a}{2D}$; $C_3 = \frac{q_\circ a^2}{24D}$

$$\frac{dw}{dr} = \frac{q_\circ r^3}{16D} + \frac{C_1 r}{2}$$

$$\frac{1}{r}\frac{d\left(r\left(\frac{q_\circ r^3}{16D} + \frac{C_1 r}{2}\right)\right)}{dr} = \frac{1}{r}\frac{d\left(\left(\frac{q_\circ r^4}{16D} + \frac{C_1 r^2}{2}\right)\right)}{dr} = \frac{1}{r}\left(\left(\frac{q_\circ r^3}{4D} + C_1 r\right)\right)$$

$$r\frac{d\left[\frac{1}{r}\left(\frac{q_\circ r^3}{4D} + C_1 r\right)\right]}{dr} = r\frac{d\left(\left(\frac{q_\circ r^2}{4D} + C_1\right)\right)}{dr} = \left(\frac{q_\circ r^2}{2D}\right)$$

$$\frac{1}{r}\frac{d\left(\frac{q_\circ r^2}{2D}\right)}{dr} = \frac{1}{r}\frac{q_\circ r}{D} = \frac{q_\circ}{D}$$

14.7.4 What is the maximum vertical deflection of the uniformly loaded plate in cylindrical bending (Fig. 14.4)?

Solution

From symmetry, the maximum deflection is at the center $x = \frac{a}{2}$. The plate deflection equation with known constants is:

$$w(x) = \frac{q_\circ x^4}{24D} - \frac{q_\circ a x^3}{12D} + \frac{q_\circ a^3}{24D}x$$

At $x = \frac{a}{2}$:

$$w\left(\frac{a}{2}\right) = \frac{q_\circ \left(\frac{a}{2}\right)^4}{24D} - \frac{q_\circ a \left(\frac{a}{2}\right)^3}{12D} + \frac{q_\circ a^3}{24D}\frac{a}{2}$$

$$w\left(\frac{a}{2}\right) = \frac{q_\circ a^4}{12D}\left(\frac{1}{32} - \frac{4}{32} + \frac{8}{32}\right) = \frac{q_\circ a^4}{12D}\left(\frac{5}{32}\right) = \frac{5q_\circ a^4}{384D}$$

Note that for negative $q_\circ$, the deflection is negative.

References

Kirchhoff, G. (1850). Journal of Mathematics (Crelle), *40*.

Libai, A., & Simmonds, J. G. (1998). *The nonlinear theory of elastic shells*. Cambridge: Cambridge University Press.

Navier, C.-L. (1823). Bull. Soc. Philomath, Paris, p. 92 (see Timoshenko, *History of strength of materials*, p. 121, 1953).

Shames, I. H., & Dym, C. L. (1985). *Energy and finite element methods in structural mechanics.* Hemisphere Publishing.

Timoshenko, S. P., & Woinowsky-Krieger, S. (1959). *Theory of plates and shells.* McGraw-Hill.

Chapter 15
Computational Mechanics

There are a variety of approximate methods available for solving problems in solid mechanics; they include, but are not limited to, approximate analytical, variational, finite difference, and finite element methods. All of these methods can be categorized under the general heading of "computational mechanics." Finite element methods are the most common and most robust; they will be introduced in the following.

15.1 Finite Element Methods

Analysis of complicated structural configurations can be accomplished through the use of *finite elements*. The fundamental idea is to approximate the structure by a series of small elemental configurations over which some feature, such as the displacement field, has a predefined variation. A simple example is to represent the structure by a series of triangular elements over which the displacements are assumed to vary linearly. Linear variation of the displacements results in strains that are constant over the element. For linear elastic, homogeneous materials, the stresses are also constant over the elements. As the number of elements increases, the approximation to the actual stress and strain distribution over the structure becomes more accurate.

The development of finite element methods grew out of the work of Courant (1943), Turner et al. (1956), Clough (1960) and Argyris and Kelsey (1960). All of these early applications were on problems in solid mechanics. Clough (1960) apparently was the first to use the term "finite element." The field has grown exponentially since that time and is now the preferred method of analysis for most structural problems as well as problems in other fields including, but certainly not limited to, fluid mechanics, heat transfer, and biomechanics. Methods are available for one-dimensional, two-dimensional, and three-dimensional linear and nonlinear problems, for both static and dynamic analyses. The method has proven to be

C.T. Herakovich, *A Concise Introduction to Elastic Solids*,
DOI 10.1007/978-3-319-45602-7_15

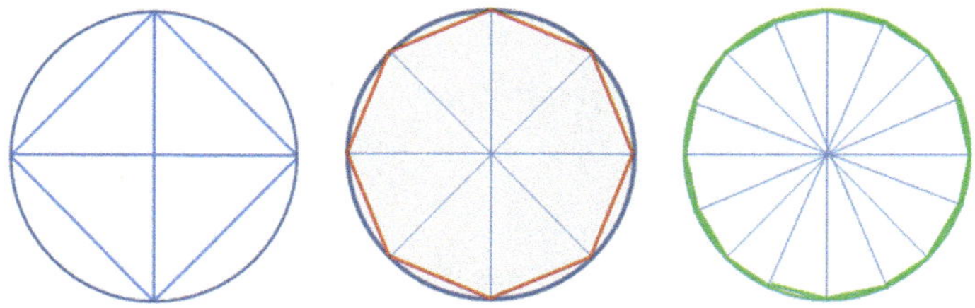

Fig. 15.1 Circle approximated by triangular finite elements

computationally efficient and robust; it can be applied to a wide variety of problems. There can be no question that the availability or powerful computers has enhanced the use of finite element methods. For an introduction to the finite element method, see Huebner (1975) or Reddy (1984).

15.2 The Value of π

One of the simplest examples of the use of finite elements is the determination of the value of π knowing that the area of a circle, A, is proportional to the square of the radius, R. While this is not an application in solid mechanics, it does demonstrate the fundamental idea behind the use of finite elements. Figure 15.1 shows an increasing number of triangular elements (N) for approximating the area of a circle, as N increases from 4 to 16. As the number of elements is increased, the error in approximating the area of the circle decreases and thus the approximate value of π becomes more accurate. Additional approximate values are shown in the figure for N = 180 and N = 360. For N = 360, the approximate value of π is 3.1414. It should be obvious from the figures that symmetry can be invoked to reduce the number of elements required for the calculation. Symmetry is used as much as possible in analyzing engineering structures in order to minimize the computational resources required.

Fig. 15.2 Square bar in torsion

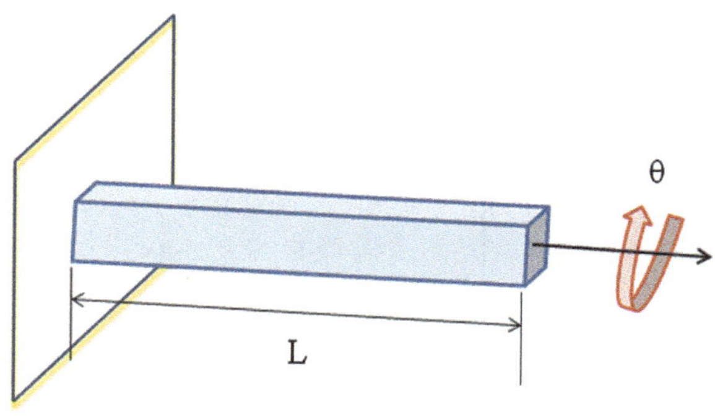

$$A = \pi R^2 \tag{15.1}$$

15.3 Torsion of a Square Bar by Finite Elements

Exact solutions for torsion of non-circular bars exist only for a limited number of cross-sectional shapes (see Timoshenko and Goodier 1934, Chap. 10). When the cross section is non-circular, approximate methods generally are used. The application of finite elements to the problem of torsion of non-circular bars demonstrates both the simplicity and the power of the method.

Consider a bar with a square cross section as shown in Fig. 15.2. Since the cross section is constant along the length of the bar, it is only necessary to investigate the distribution of stresses over a generic cross section. Boundary conditions require the shear stresses be zero at the corners. A gradation in the size of the elements is used to capture this variation in stresses as the corners is approached (Fig. 15.3).

Prandtl (1903) has shown that the nonzero shear stresses in a solid, non-circular bar subjected to torsion can be represented in terms of an unknown stress function $\varphi(x, y)$ such that:

$$\tau_{zx} = \frac{\partial \varphi}{\partial y} \quad \text{and} \quad \tau_{zy} = \frac{\partial \varphi}{\partial x} \tag{15.2}$$

The solution is then the function $\varphi(x, y)$ that satisfies:

$$\frac{\partial^2 \varphi}{\partial x^2} + \frac{\partial^2 \varphi}{\partial y^2} + 2G\theta = 0 \tag{15.3}$$

with $\varphi = 0$ on the boundary. In the above, G is the shear modulus of the material and θ is the angle of twist per unit length of bar.

For the finite element formulation, the cross section is represented by a system of triangular finite elements as shown in Fig. 15.3, where symmetry has been invoked to reduce the analysis to one-eighth of the cross section. The finite element solution

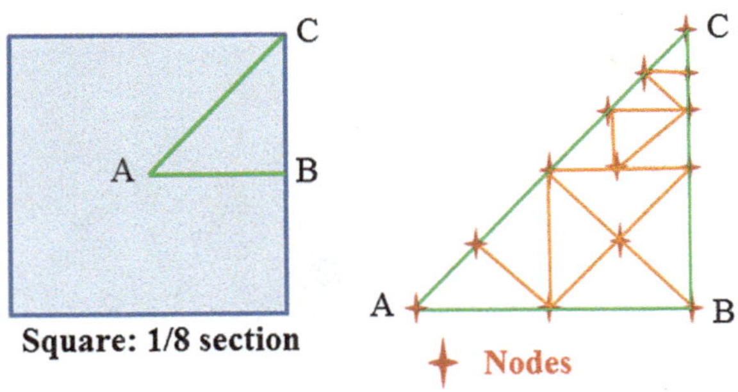

Fig. 15.3 Finite element representation of a square cross section

method assigns unknown nodal values φ_i to each node (see figure) in the finite element representation. The method then consists in establishing a system of simultaneous algebraic equations to determine the unknown φ_i values.

There are, in general, two approaches for determining the system of simultaneous equations. One method is referred to as the *direct method* and the other is a *variational method.* The variational method is more general. It is based upon minimization of a functional, such as the potential or complementary energy (see Fung (1965), Sects. 10.7 and 10.9), defined over the region of interest. Partial derivatives of this functional, with respect to the unknown nodal values, are set to zero to provide a system of simultaneous equations for determination of the nodal values. The number of equations is equal to the number of unknowns. With the nodal values determined and the specified linear variation of φ over the elements, the shear stresses are constant over each element. Thus, the problem is solved.

15.4 Exercises

15.4.1 Show that the value for π in Fig. 5.1 for 180 triangular elements is correct.
15.4.2 Show that the value for π in Fig. 5.1 for 360 triangular elements is correct.
15.4.3 Estimate the value of π using a circumscribed square.
15.4.4 Estimate the value of π using a circle circumscribed about a square.
15.4.5 Show that if it assumed the stress function ϕ varies linearly over each triangular element in a finite element representation of the torsion problem, this is equivalent to assuming that the shear stresses are constant over each element.

Appendix: Solutions

15.4.1 Show that the value for π in Fig. 15.1 for 180 triangular elements is correct.

<u>**Solution**</u>

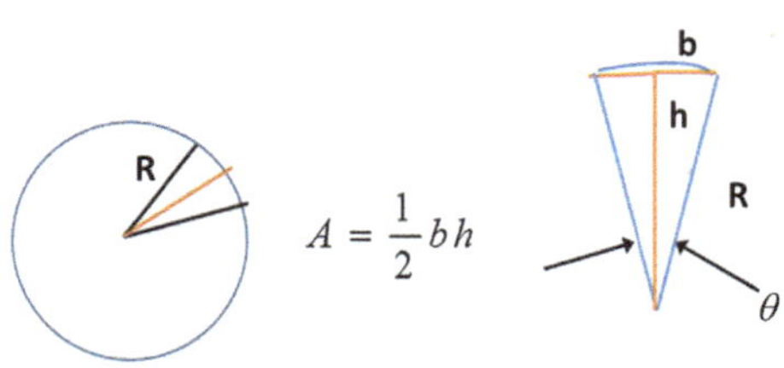

$$b = R \sin\left(\frac{\theta}{2}\right) \quad h = R \cos\left(\frac{\theta}{2}\right)$$

$$A = \frac{1}{2} R^2 \sin\left(\frac{\theta}{2}\right) \cos\left(\frac{\theta}{2}\right)$$

See table below.

15.4.2 Show that the value for π in Fig. 5.1 for 360 triangular elements is correct.

Solution

The results below are from an Excel file.

N	theta	rads	rads/2	sin(t/2)	cos(t/2)	1/2(s t/2 c t/2)	pi
4.00	90.00	1.570796	0.785398	0.707107	0.707107	0.500000	2.000000
8.00	45.00	0.785398	0.392699	0.382683	0.923880	0.353553	2.828427
16.00	22.50	0.392699	0.196350	0.195090	0.980785	0.191342	3.061467
180.00	2.00	0.034907	0.017453	0.017452	0.999848	0.017450	3.140955
360.00	1.00	0.017453	0.008727	0.008727	0.999962	0.008726	3.141433

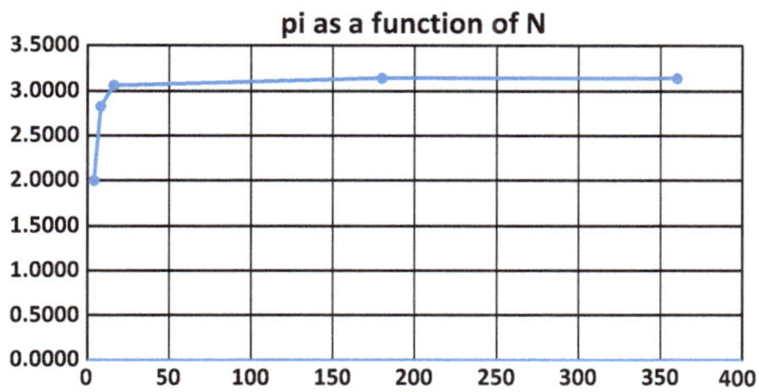

15.4.3 Estimate the value of π using a circumscribed square.

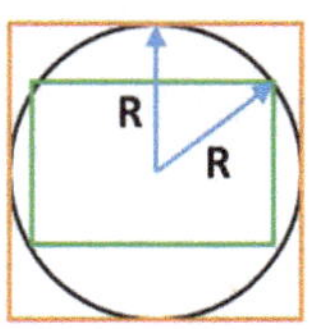

$$A_{red} = (2R)(2R) = 4R^2 \quad \Rightarrow \pi = 4$$

15.4.4 Estimate the value of π using a circle circumscribed about a square.

Solution

$$A_{green} = [2R\,\sin(45)]^2 = 4R^2\left(\frac{1}{\sqrt{2}}\right)^2 = 2R^2 \quad \pi = 2$$

15.4.5 Show that if it assumed the stress function φ varies linearly over each triangular element in a finite element representation of the torsion problem, this is equivalent to assuming that the shear stresses are constant over each element.

Solution

From (15.2):

$$\tau_{zx} = \frac{\partial \varphi}{\partial y} \text{ and } \tau_{zy} = \frac{\partial \varphi}{\partial x}$$

For linear variation of ϕ:

$$\phi = Ax + By$$

$$\frac{\partial \phi}{\partial x} = A \quad \frac{\partial \phi}{\partial y} = B$$

Thus, the shear stresses are constant over each finite element.

References

Argyris, J. H., & Kelsey, S. (1960). *Energy theorems and structural analysis*. Butterworth Scientific.

Clough, R. W. (1960). *The finite element method in plane stress analysis*. In *Proceedings of 2nd conference on Journal of the Structural Division*. Electronic Computation.

Courant, R. (1943). Variational methods for the solution of problems of equilibrium and vibrations. *Bulletin of the American Mathematical Society, 49*, 1–43.

Fung, Y. C. (1965). *Foundations of solid mechanics*. Upper Saddle River: Prentice-Hall.

Huebner, K. H. (1975). *The finite element method for engineers*. New York: Wiley.

Prandtl, L. (1903). Zur torsion von prismatischen stäben. *Phys Zeitschr, 4*, 758–770.

Reddy, J. N. (1984). *An introduction to the finite element method*. New York: McGraw-Hill.

Timoshenko, S. P., & Goodier, J. N. (1934). *Theory of elasticity*. New York: McGraw-Hill.

Turner, M., Clough, R., Martin, H., & Topp, L. (1956). Stiffness and deflection analysis of complex structures. *Journal of Aerosol Science, 23*, 805–823.

Chapter 16
Fibrous Composite Materials

The use of fibrous composite materials for structural applications became quite common as the twentieth century came to a close. The confidence in and ability to design and manufacture with these materials was exemplified by the twenty-first-century application of carbon fiber composite as the primary load-bearing structural material in the Boeing 787 commercial aircraft (Fig. 16.1). Glass fiber composites are also in extensive use. The advantages of fibrous composites include high stiffness, high strength, light weight, non-corrosive, zero or near-zero thermal expansion, exceptional fatigue life, high impact resistance, reduced manufacturing costs due to reduced parts count and reduced maintenance costs.

Table 16.1 (from Herakovich 1998) presents material properties for a variety of metals, unidirectional fibers, and matrix materials. Potential weight savings with composites are demonstrated through the ratios of the specific stiffness and specific strength to those of aluminum. The actual properties of composites will not be as high as those for fibers alone because the composite is a combination of fiber and matrix, and off-axis layers reduce many of the properties.

The properties of fibrous composite materials vary with direction. This results in more complex constitutive equations and analysis methods. Materials whose properties vary with direction are called anisotropic materials. While analysis and design with composites are more involved with isotropic materials, the availability of modern computers renders the problem very manageable.

A historical review of the developments in mechanics of composites is given by Herakovich (2012).

C.T. Herakovich, *A Concise Introduction to Elastic Solids*,
DOI 10.1007/978-3-319-45602-7_16

Fig. 16.1 Boeing 787

Table 16.1 Properties of engineering materials, fibers, and matrix

Material	Density ρ	Modulus EL	Poisson's ratio ν_L	Strength σuL	Specific stiffness (E/ρ)/(E/ρ)Al	Specific strength (σu/ρ)/(σu/ρ)Al	Thermal expansion coefficient αL
	g/cm^3 (lb/in^3)	GPa (Msi)		MPa (ksi)			μ/°C (μ/°F)
Metals							
Steel	7.8 (0.284)	200 (29)	0.32	1724 (250)	1.0	1.2	12.8 (7.1)
Aluminum	2.7 (0.097)	69 (10)	0.33	483 (70)	1.0	1.0	23.4 (13.0)
Titanium	4.5 (0.163)	91 (13.2)	0.36	758 (110)	0.95	1.2	8.8 (4.9)
Fibers (axial properties)							
AS4	1.80 (0.065)	235 (34))	0.20	3599 (522)	5.1	11.1	−0.8 (−0.44)
T300	1.76 (0.064)	231 (33)	0.20	3654 (530)	5.1	11.5	−0.5 (−0.3)
P100S	2.15 (0.078)	724 (105)	0.20	2199 (319)	13.2	5.5	−1.4 (−0.78)
IM8	1.8 (0.065)	310 (45)	0.20	5171 (750)	6.7	16.1	–
Boron	2.6 (0.094)	385 (55.8)	0.21	3799 (551)	5.8	8.3	8.3 (4.6)
Kevlar 49	1.44 (0.052)	124 (18)	0.34	3620 (525)	3.6	13.9	−2.0 (−1.1)
SCS-6	3.3 (0.119)	400 (58.0)	0.25	3496 (507)	5.1	6.1	5.0 (2.77)
Alumina	3.95 (0.143)	379 (55)	0.25	1585 (230)	3.7	1.9	7.5 (4.2)
S-2 Glass	2.46 (0.090)	86.8 (12.6)	0.23	4585 (665)	1.4	10.4	1.6 (0.9))
E-Glass	2.58 (0.093)	69 (10.0)	0.22	3450 (550)	1.05	7.5	5.4 (3.0)
Matrix materials							
Epoxy	1.38 (0.050)	4.6 (0.67)	0.36	58.6 (8.5)	0.08	0.4	63 (35)
Polyimide	1.46 (0.053)	3.5 (0.5)	0.35	103 (15)	0.03	0.4	36 (20)
Copper	8.9 (0.32)	117 (17)	0.33	400 (58)	0.5	0.3	17 (9.4)
Silicon carbide	3.2 (0.116)	400 (58)	0.25	310 (45)	4.9	0.5	4.8 (2.67)

16.1 Laminated Composites

Assemblages of layers (Fig. 16.2) of fibrous composite materials, which can be tailored to provide a wide range of engineering properties, including inplane stiffness, bending stiffness, strength, and coefficients of thermal expansion, are called composite laminates.

The individual layers consist of high-modulus, high-strength fibers embedded in a polymeric, metallic, or ceramic matrix material. Fibers currently in use include carbon, glass, aramid, boron, and silicon carbide. Matrix materials that are in use include thermoplastic and thermoset resins, ceramic, and metallic (see Table 16.1).

Layers of different materials may be used, resulting in a *hybrid* laminate. The individual layers of a laminate generally are orthotropic (principal properties in orthogonal directions) or transversely isotropic (isotropic properties in a transverse plane of the layer). Laminates may exhibit anisotropic (variable direction of principal properties), orthotropic, or quasi-isotropic properties. Quasi-isotropic laminates exhibit isotropic (independent of direction) inplane response, but do not exhibit isotropic out-of-plane (bending) response. Depending upon the stacking sequence of the individual layers, the laminate may exhibit coupling between inplane and out-of-plane responses. An example of bending–stretching coupling is the presence of curvature developing as a result of inplane loading. Likewise, inplane strains may develop as a result of pure moment loading for an unsymmetric laminate.

Fig. 16.2 Laminated composite

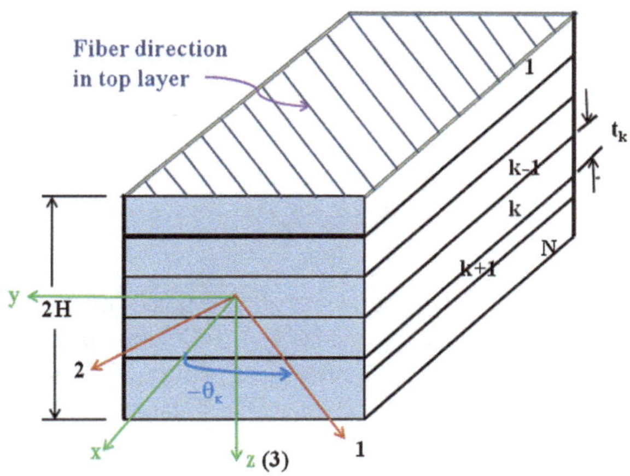

16.2 Plane Stress of Orthotropic Material

A layer of a unidirectional, fibrous composite can be represented as a homogeneous, orthotropic material with effective properties in the 1, 2, and 3 directions. For plane stress in the 1–2 plane, and using the reduced notation of composites, the constitutive equations can be written:

$$\begin{aligned} \varepsilon_1 &= \frac{\sigma_1}{E_1} - \nu_{21}\frac{\sigma_2}{E_2} \\ \varepsilon_2 &= -\nu_{12}\frac{\sigma_1}{E_1} + \frac{\sigma_2}{E_2} \\ \gamma_{12} &= \frac{\tau_{12}}{G_{12}} \end{aligned} \tag{16.1}$$

This constitutive equation can be written in matrix form as:

$$\begin{Bmatrix} \varepsilon_1 \\ \varepsilon_2 \\ \gamma_{12} \end{Bmatrix} = \begin{bmatrix} \frac{1}{E_1} & -\frac{\nu_{21}}{E_2} & 0 \\ -\frac{\nu_{12}}{E_1} & \frac{1}{E_2} & 0 \\ 0 & 0 & \frac{1}{G_{12}} \end{bmatrix} \begin{Bmatrix} \sigma_1 \\ \sigma_2 \\ \tau_{12} \end{Bmatrix} \tag{16.2}$$

In (16.2) E_1 is the modulus in the 1 (fiber) direction, E_2 is the modulus in the 2 (transverse) direction, G_{12} is the shear modulus in the 1–2 plane, ν_{12} is the Poisson's ratio for loading in the 1-direction, and ν_{21} is the Poisson's ratio for loading in the 2-direction.

The 3 × 3 matrix in brackets in Eq. (16.2) is the stiffness matrix $[Q]$ for a unidirectional layer. It can be transformed to other orientations where it is identified as $[\bar{Q}]$.

16.3 Classical Lamination Theory

Lamination theory describes the response of a composite laminate subjected to inplane and bending loads. The following presentation closely follows that of Herakovich (1998). The laminate in Fig. 16.2 employs a global x–y–z coordinate system with z perpendicular to the plane of the laminate and positive downward. The origin of the coordinate system is located on the laminate midplane. The laminate has N layers, numbered from top to bottom. Each layer has a distinct fiber orientation θ_k. The z coordinate to the bottom of the kth layer is designated z_k with the top of the layer being at z_{k-1}. The thickness, t_k, of any layer is then

$t_k = z_k - z_{k-1}$. The top surface of the laminate is denoted z_0 and the total thickness is $2H$.

It is assumed that each layer can be represented as a homogeneous material with known effective properties which may be isotropic, orthotropic, or transversely isotropic and that there is perfect bonding between layers. Further, it is assumed that each layer is in a state of plane stress and the laminate deforms according to the Kirchhoff (1850) assumptions for bending and stretching of thin plates. According to this theory, planes normal to the midplane prior to bending remain straight and normal to the deformed midplane after bending, and lines normal to the midplane do not change length.

Prior to delving into the details of stress analysis for laminated composite materials, we note that it is common practice to employ a reduced notation for most quantities of interest. Most notably, single subscripts are used rather than double subscripts when it is obvious as to intent. For example, we define $\sigma_{xx} \equiv \sigma_x$ and $\varepsilon_{xx} \equiv \varepsilon_x$. We also use matrix or tensor notation to simplify the writing of otherwise complex quantities. For example:

$$\{\sigma\} \equiv \begin{Bmatrix} \sigma_x \\ \sigma_y \\ \tau_{xy} \end{Bmatrix} \tag{16.3}$$

$$\{\varepsilon\} \equiv \begin{Bmatrix} \varepsilon_x \\ \varepsilon_y \\ \gamma_{xy} \end{Bmatrix} \tag{16.4}$$

The Kirchhoff assumptions require that the out-of-plane strains ε_z, γ_{zy} and γ_{zx} be identically zero. The inplane strains, $\{\varepsilon\} = \begin{Bmatrix} \varepsilon_x \\ \varepsilon_y \\ \gamma_{xy} \end{Bmatrix}$, at any z location, can be expressed in terms of the *midplane strains*, $\{\varepsilon^\circ\}$:

$$\{\varepsilon^\circ\} = \begin{Bmatrix} \varepsilon_x^\circ \\ \varepsilon_y^\circ \\ \gamma_{xy}^\circ \end{Bmatrix} \tag{16.5}$$

and the curvatures, $\{\kappa\}$

$$\{\kappa\} = \begin{Bmatrix} \kappa_x \\ \kappa_y \\ \kappa_{xy} \end{Bmatrix} \tag{16.6}$$

as:

$$\{\varepsilon\} = \{\varepsilon^{\circ}\} + z\{\kappa\} \tag{16.7}$$

The stresses, $\{\sigma\}^k$, at any z location in the kth layer are determined from these strains and the layer constitutive equation expressed in terms of the stiffness matrix $[\bar{Q}]^k$ of the kth layer. The resulting constitutive equation can be written as:

$$\{\sigma\}^k = [\bar{Q}]^k(\{\varepsilon^{\circ}\} + z\{\kappa\}) \tag{16.8}$$

The stiffness matrix $[\bar{Q}]$ is a function of the engineering properties (modulii, Poisson ratios, and shear moduli) of the layer and the orientation of the layer. (See Herakovich (1998) for details.) The stiffness matrix is symmetric and has the general form:

$$[\bar{Q}] = \begin{bmatrix} \bar{Q}_{11} & \bar{Q}_{12} & \bar{Q}_{16} \\ \bar{Q}_{12} & \bar{Q}_{22} & \bar{Q}_{26} \\ \bar{Q}_{16} & \bar{Q}_{26} & \bar{Q}_{66} \end{bmatrix} \tag{16.9}$$

In the absence of curvature ($\kappa = 0$), the expanded plane stress constitutive equation for the kth layer has the form:

$$\begin{Bmatrix} \sigma_x \\ \sigma_y \\ \tau_{xy} \end{Bmatrix}^k = \begin{bmatrix} \bar{Q}_{11} & \bar{Q}_{12} & \bar{Q}_{16} \\ \bar{Q}_{12} & \bar{Q}_{22} & \bar{Q}_{26} \\ \bar{Q}_{16} & \bar{Q}_{26} & \bar{Q}_{66} \end{bmatrix}^k \begin{Bmatrix} \overset{\circ}{\varepsilon}_x \\ \overset{\circ}{\varepsilon}_y \\ \overset{\circ}{\gamma}_{xy} \end{Bmatrix} \tag{16.10}$$

We note that the 16 and 26 subscript notation (as opposed to 13 and 23) is used in composite laminates to identify shear-related terms from the overall 6×6 matrix that is required for a fully anisotropic material.

The inplane forces per unit length $\{N\}$ (Fig. 16.3) are defined as the through-the-thickness integrals of the planar stresses in the laminate.

$$\{N\} = \int_{-H}^{H} \{\sigma\} dz \tag{16.11}$$

Combining (16.8) with (16.9) and integrating give the general form of the relationship for inplane forces in terms of midplane strains, curvatures, and laminate parameters $[A]$ $[B]$ defined below:

$$\{N\} = [A]\{\varepsilon^{\circ}\} + [B]\{\kappa\} \tag{16.12}$$

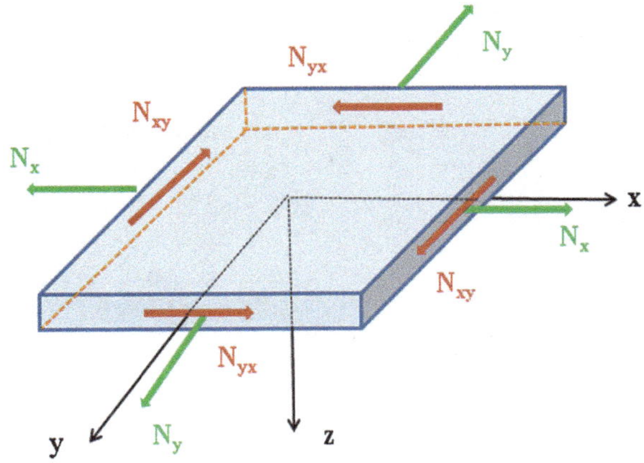

Fig. 16.3 Inplane forces

The $[A]$ and $[B]$ are coefficient matrices that are functions of the layer material properties and laminate stacking sequence. The $[A]$ matrix represents the inplane stiffness, and the $[B]$ matrix defines the bending–stretching coupling. $[A]$ is a function of the layer thicknesses but is independent of the stacking sequence of the layers. In contrast, $[B]$ is dependent on both the layer thicknesses and the stacking sequence of the individual layers. For a laminate whose stacking sequence is symmetric about the laminate midplane, $[B] \equiv 0$, and there is no bending–stretching coupling. The material properties, stiffnesses $[\bar{Q}]$, are constant in each layer. Thus, the integration can be replaced as a summation over all the layers. The explicit expressions for [A] and [B] are:

$$\begin{aligned} [A] &= \sum_{k=1}^{N} [\bar{Q}]^k (z_k - z_{k-1}) \\ [B] &= \frac{1}{2} \sum_{k=1}^{N} [\bar{Q}]^k \left(z_k^2 - z_{k-1}^2\right) \end{aligned} \tag{16.13}$$

The moments per unit length, $\{M\}$ (Fig. 16.4), are defined as the integrals of the differential moments, $\{\sigma\}zdz$ integrated over the laminate thickness

$$\{M\} = \int_{-H}^{H} \{\sigma\} z dz \tag{16.14}$$

Fig. 16.4 Bending moments

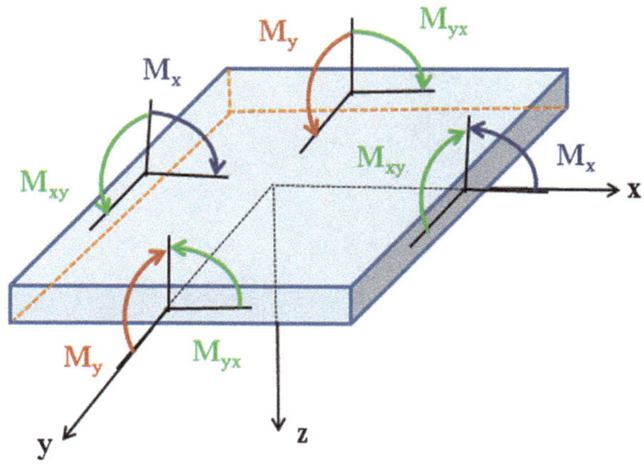

Introducing Eq. (16.8) for stresses in (16.14) and integrating give the moment equations in the form

$$\{M\} = [B]\{\varepsilon^{\circ}\} + [D]\{\kappa\} \tag{16.15}$$

Here, the bending stiffness matrix, $[D]$, is a coefficient matrix similar to $[A]$ and $[B]$. It is stacking sequence dependent and not identically zero for a symmetric laminate. The explicit expression for [D] is:

$$[D] = \frac{1}{3}\sum_{k=1}^{N} [\bar{Q}]^k \left(z_k^3 - z_{k-1}^3\right) \tag{16.16}$$

Combining the force and moment equations into a single matrix form gives the *fundamental equation of lamination theory.*

$$\begin{Bmatrix} N \\ M \end{Bmatrix} = \begin{bmatrix} A & B \\ B & D \end{bmatrix} \begin{Bmatrix} \varepsilon^{\circ} \\ \kappa \end{Bmatrix} \tag{16.17}$$

This equation exhibits coupling between the bending and stretching responses of a laminate through the $[B]$ matrix. If $[B] \equiv 0$, the inplane (normal and shear) response is decoupled from the bending response. Indeed, there is no bending–stretching coupling for laminates which are symmetric about their midplane since, as indicated previously, $[B] \equiv 0$ for symmetric laminates. Thus, for symmetric laminates:

$$\{N\} = [A]\{\varepsilon^{o}\} \tag{16.18}$$

$$\{M\} = [D]\{\kappa\} \tag{16.19}$$

16.4 Effective Laminate Properties

Prediction of the effective, inplane, engineering properties of symmetric, composite laminates can be determined from thought experiments by starting with the force–midplane strain Eq. (16.18). For effective properties of the laminate, we want to work with average stress over the laminate thickness. Thus, we note that the *laminate average stress* $\{\bar{\sigma}\}$ is given by:

$$\{\bar{\sigma}\} \equiv \frac{\{N\}}{2H} \tag{16.20}$$

Combining (16.18) and (16.20) gives an expression for the midplane strains in terms of the laminate averages stress:

$$\{\varepsilon^o\} = 2H[A]^{-1}\{\bar{\sigma}\} \tag{16.21}$$

In (16.21), $[A]^{-1}$ is the inverse of $[A]$. For convenience, we define

$$2H[A]^{-1} \equiv [a^*] \tag{16.22}$$

$[a^*]$ is called the *laminate compliance*. We can now write the fundamental effective stress–strain relationship for symmetric, composite laminates in terms of the compliance:

$$\{\varepsilon^{\circ}\} = [a^*]\{\bar{\sigma}\} \tag{16.23}$$

Equation (16.23) is the *fundamental effective stress–strain relationship for a symmetric, composite laminate*. We can use this equation for thought experiments, for specified states of stress, to predict the resulting strain and, thereby, the effective laminate properties, i.e., the engineering constants.

The compliance matrix $[a^*]$ is generally a fully populated, symmetric matrix of the form:

$$[a^*] = \begin{bmatrix} a_{11}^* & a_{12}^* & a_{16}^* \\ a_{12}^* & a_{22}^* & a_{26}^* \\ a_{16}^* & a_{26}^* & a_{66}^* \end{bmatrix} \tag{16.24}$$

The fundamental stress–strain relationship (16.23) can now be expressed in expanded form as:

$$\begin{Bmatrix} \varepsilon_x^{\circ} \\ \varepsilon_y^{\circ} \\ \gamma_{xy}^{\circ} \end{Bmatrix} = \begin{bmatrix} a_{11}^* & a_{12}^* & a_{16}^* \\ a_{12}^* & a_{22}^* & a_{26}^* \\ a_{16}^* & a_{26}^* & a_{66}^* \end{bmatrix} \begin{Bmatrix} \bar{\sigma}_x \\ \bar{\sigma}_y \\ \bar{\tau}_{xy} \end{Bmatrix} \tag{16.25}$$

Examples demonstrating the variability of the effective engineering constants as a function of fiber orientations and laminate stacking sequence for unidirectional, off-axis and symmetric, angle-ply laminates are presented in the following for a specific carbon/epoxy composite. A unidirectional laminate consists of layers all having the same fiber orientation and material properties (Fig. 16.5a). A symmetric, angle-ply laminate has an equal number of layers at $+\theta$ and $-\theta$ fiber orientations located symmetrically about the laminate midplane (Fig. 16.5b).

The results in Figs. 16.6, 16.7, 16.8, and 16.9 demonstrate one of the major advantages of fibrous composite materials, namely the ability to tailor the material properties through the choice of fiber orientations, layer thicknesses, stacking

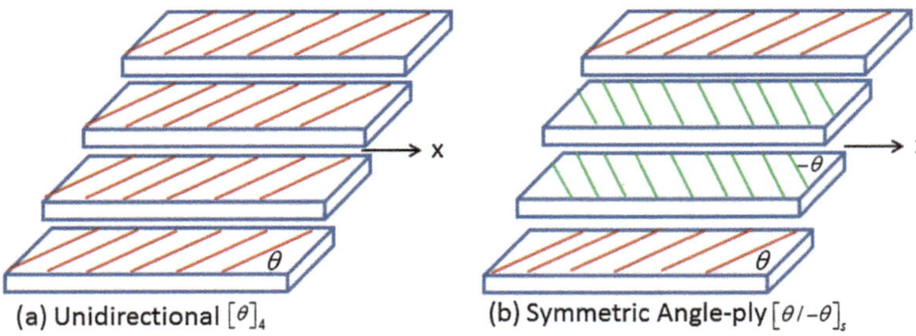

Fig. 16.5 Unidirectional and angle-ply laminates. **a** Unidirectional $[\theta]_4$. **b** Symmetric angle-ply $[\theta/-\theta]_4$

Fig. 16.6 Effective axial modulus

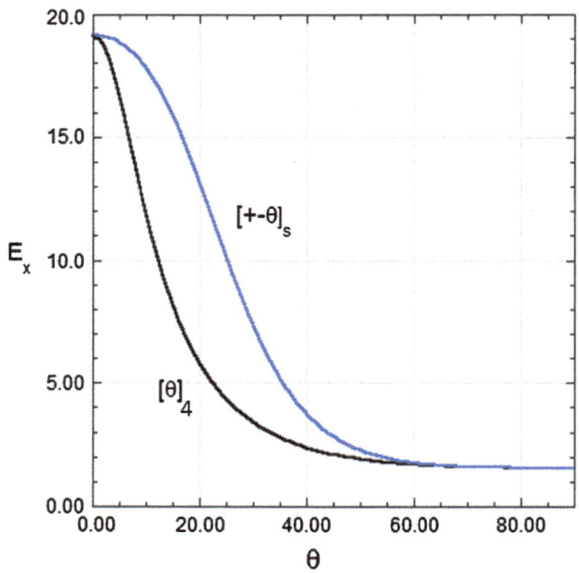

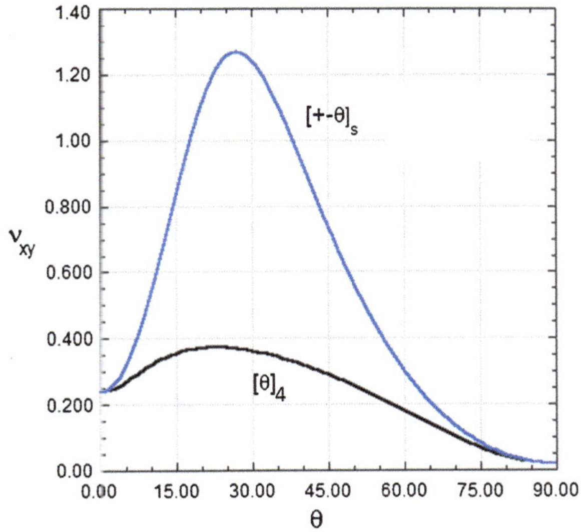

Fig. 16.7 Effective Poisson's ratio

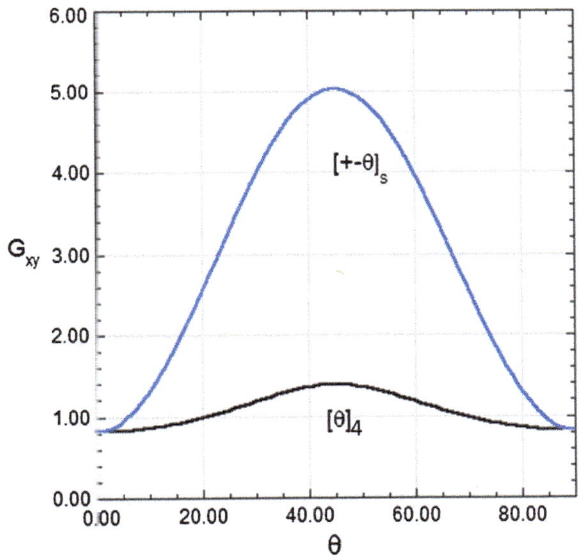

Fig. 16.8 Effective shear modulus

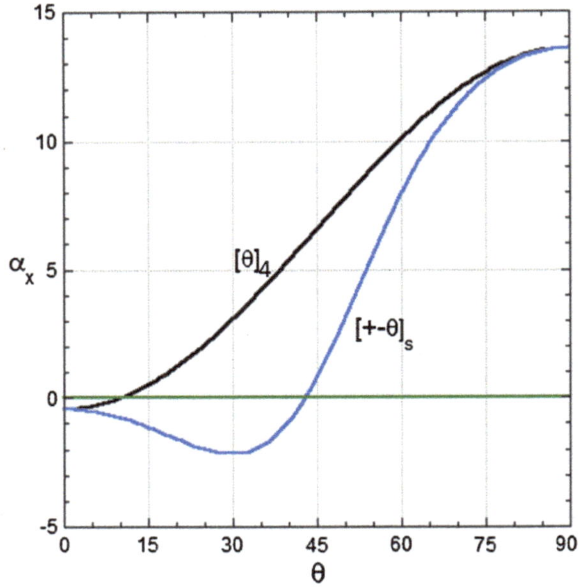

Fig. 16.9 Effective coefficient of thermal expansion

sequence, and material properties of the individual layers. The results presented here are for T300/5208 carbon/epoxy composite. This composite consists of T300 carbon fibers in an epoxy matrix (5208). The theoretical predictions have been verified by experimental results.

16.5 Effective Axial Modulus

For a uniaxial tension test, the applied state of stress is:

$$\begin{aligned}\bar{\sigma}_x &> 0\\ \bar{\sigma}_y &= 0\\ \bar{\tau}_{xy} &= 0\end{aligned} \tag{16.26}$$

Combining (16.25) and (16.26), the strains are:

$$\begin{aligned}\varepsilon_x^{\circ} &= a_{11}^{*}\bar{\sigma}_x\\ \varepsilon_y^{\circ} &= a_{12}^{*}\bar{\sigma}_x\\ \gamma_{xy}^{\circ} &= a_{16}^{*}\bar{\sigma}_x\end{aligned} \tag{16.27}$$

By definition, we then have the effective axial modulus E_x:

$$E_x = \frac{\bar{\sigma}_x}{\varepsilon_x^\circ} = \frac{1}{a_{11}^*} \tag{16.28}$$

Predictions for the effective axial modulus, E_x, of unidirectional and angle-ply laminates are presented in Fig. 16.6. The angle-ply laminate exhibits higher stiffness than the off-axis lamina for fiber angles ranging from 0° to approximately 60°. The higher axial stiffness for the angle-ply laminate is a result of the constraining effect (and resulting multiaxial state of stress) that the adjacent layers of the laminate have on each other.

16.6 Effective Inplane Poisson's Ratio

The effective Poisson's ratio ν_{xy} is defined:

$$\nu_{xy} = -\frac{\varepsilon_y^\circ}{\varepsilon_x^\circ} = -\frac{a_{12}^*}{a_{11}^*} \tag{16.29}$$

The comparison of Poisson's ratios (Fig. 16.7) is one of the most interesting and surprising results for laminated fibrous composites. There is a very large increase in Poisson's ratio when an off-axis ply is laminated with another off-axis ply of opposite sign. This is the case for a wide range of fiber orientations. The angle-ply laminate exhibits a maximum Poisson's ratio in excess of 1.25 (for this particular carbon/epoxy) at an angle of approximately 27°. That the Poisson's ratio can be in excess of 1.0 is very surprising to those accustomed to working with metals, where the maximum Poisson's ratio is 0.5.

16.7 Effective Shear Modulus

Shear modulus is a measure of the materials resistance (stiffness) to shear loading N_{xy} (Fig. 16.3). For a symmetric laminate subjected to pure shear stress $\bar{\tau}_{xy}$, the shear strain is determined from (16.25) as:

$$\gamma_{xy}^\circ = a_{66}^* \bar{\tau}_{xy} \tag{16.30}$$

By definition, the effective shear modulus, G_{xy}, is:

$$G_{xy} = \frac{\bar{\tau}_{xy}}{\gamma_{xy}^{\circ}} = \frac{1}{a_{66}^{*}} \tag{16.31}$$

The angle-ply laminate is much stiffer than the unidirectional lamina (Fig. 16.8) for essentially all fiber orientations. At $\theta = 45°$, where the shear stiffness of both the lamina and the laminate are largest, the stiffness of the laminate is more than 3.5 times that of the lamina for the carbon/epoxy under consideration. The results clearly indicate that $\pm 45°$ fiber orientations are desired in structures requiring high shear stiffness.

16.8 Effective Coefficient of Thermal Expansion

The fundamentals of thermal expansion for isotropic materials were discussed in Chap. 11. If a material is isotropic, there is a single coefficient of thermal expansion, α, that influences the thermal behavior. Most materials have positive *coefficients of expansion* (CTE) and thus expand when heated and contract when cooled. Fibrous composite materials are not isotropic and, hence, the coefficient of thermal expansion is direction dependent.

A unidirectional fibrous composite is an orthotropic material with three distinct coefficients of thermal expansion, one in each of the three principal material directions. For the composite, we express the coefficients as a matrix, $\{\alpha\}$, where:

$$\{\alpha\} \equiv \begin{Bmatrix} \alpha_1 \\ \alpha_2 \\ \alpha_3 \end{Bmatrix} \tag{16.32}$$

When unidirectional plies are combined to form a laminate, the *effective coefficients of thermal expansion* are a function of the laminate configuration (i.e., the ply thicknesses and ply orientations) and the direction of interest. For laminates, we are interested primarily in the effective, inplane coefficients of thermal expansion expressed as:

$$\{\alpha\} = \begin{Bmatrix} \alpha_x \\ \alpha_y \\ \alpha_{xy} \end{Bmatrix} \tag{16.33}$$

Note that shear strain may be present for all directions.

The axial (fiber direction) coefficient of thermal expansion, α_1, of a unidirectional composite can be positive, negative, or zero. As will be shown, this has a profound impact on the possible effective, inplane coefficients of thermal expansion of laminates; indeed, they may be very small and they may be negative. Laminates with low or zero coefficient of thermal expansion are particularly important because they do not expand or contract when exposed to a temperature change. This can be a very important design consideration for applications where there are variations in temperature over time or where the temperature varies over the regions of the structure.

We note that when unidirectional layers are combined to form a laminate, thermal expansion is no longer "free thermal expansion." The layers of the laminate are (assumed) to be perfectly bonded. Layers at different fiber orientations constrain the adjacent layers and an internal state of stress is initiated.

Effective coefficients of thermal expansion (CTE), $\{\bar{\alpha}\}$ for laminates can be predicted starting from the expression for the *total thermal–mechanical strains*, $\{\varepsilon\}$, written as a superposition of the strains associated with stress $\{\varepsilon^\sigma\}$ and the "free" thermal strains $\{\varepsilon^T\}$:

$$\{\varepsilon\} = \{\varepsilon^\circ\} + z\{\kappa\} = \{\varepsilon^\sigma\} + \{\varepsilon^T\} \tag{16.34}$$

The layer constitutive equation gives the stress in terms of the "stress-related" strain as:

$$\{\sigma\} = [\bar{Q}]\{\varepsilon^\sigma\} \tag{16.35}$$

Combining (16.34 and 16.35) gives the fundamental equation for the stresses in a layer of a laminate that is subjected to thermal loading:

$$\{\sigma\} = [\bar{Q}]\left(\{\varepsilon^\circ\} + z\{\kappa\} - \{\varepsilon^T\}\right) \tag{16.36}$$

For a symmetric laminate, the curvature $\{\kappa\} = 0$ and the total strain in each layer is the midplane strain $\{\varepsilon^\circ\}$. We now define the *laminate, effective coefficient of thermal expansion* $\{\bar{\alpha}\}$ as:

$$\{\bar{\alpha}\} = \frac{\{\varepsilon^\circ\}}{\Delta T} \tag{16.37}$$

The problem has now been shown to be equivalent to determination of the midplane strain $\{\varepsilon^\circ\}$ for a symmetric laminate that is subjected a change in temperature—and no other external loads.

For a symmetric laminate with no external applied moments, the curvature and external force $\{N\}$ are zero. The external force is the integral of stresses over the laminate thickness. Thus, using (16.36) with $\{\kappa\} = 0$, we have:

$$\{N\} = 0 = \int_{-H}^{H} [\bar{Q}]^k \left(\{\varepsilon^\circ\} - \{\alpha\}^k \Delta T\right) dz \tag{16.38}$$

With $\{\varepsilon^\circ\}$ a constant and using the definition of $[A]$, (16.38) reduces to:

$$[A]\{\varepsilon^\circ\} = \Delta T \int_{-H}^{H} [\bar{Q}]^k \{\alpha\}^k dz \tag{16.39}$$

Solving for the midplane strain and using the definition (16.37) for the laminate coefficient of thermal expansion, we have:

$$\{\bar{\alpha}\} = [A]^{-1} \int_{-H}^{H} [\bar{Q}]^k \{\alpha\}^k dz \tag{16.40}$$

Further, since all quantities are constant within each layer, we can write:

$$\{\bar{\alpha}\} = [A]^{-1} \sum_{k=1}^{N} [\bar{Q}]^k \{\alpha\}^k t^k \tag{16.41}$$

While this expression may seem involved at first glance, the use of modern-day computers renders the solution to a straight forward procedure.

Axial CTE predictions for unidirectional and angle-ply laminates of T300/5208, as a function of fiber orientation, are presented in Fig. 16.9. As indicated in the figure, the CTE for the unidirectional lamina is a small negative value at $\theta = 0^\circ$, passes through zero at approximately 10° and attains a maximum positive value at 90°. More interesting are the results for angle-ply laminates. The CTE of the laminate decreases further from the small negative value at $\theta = 0^\circ$. As the fiber orientation increases, the CTE attains its largest negative value at approximately 30°, passes through zero at approximately 42°, and ends at the high positive value for $\theta = 90^\circ$.

16.9 Effective Coefficient of Mutual Influence

In order to complete the set of effective material properties for laminated composites, we define a new material property, the coefficient of mutual influence, $\eta_{xy,x}$; it is the ratio of the (possible) shear strain associated with axial strain:

$$\eta_{xy,x} = \frac{\gamma_{xy}^{\circ}}{\varepsilon_x^{\circ}} = \frac{a_{16}^*}{a_{11}^*} \tag{16.42}$$

16.10 Exercises

16.10.1 Confirm Eq. (16.10).
16.10.2 Confirm the expression for the inplane stiffness $[A]$ in Eq. (16.13).
16.10.3 Confirm the expression for the bending–stretching coupling $[B]$ in Eq. (16.13).
16.10.4 Confirm the expression for the bending stiffness $[D]$ in Eq. (16.16).

Appendix: Solutions

16.10.1 Confirm Eq. (16.10)

Solution
The constitutive Eq. (16.8) is:

$$\{\sigma\}^k = [\bar{Q}]^k(\{\varepsilon^{\circ}\} + z\{\kappa\})$$

For zero curvature with expanded form, this becomes:

$$\begin{Bmatrix} \sigma_x \\ \sigma_y \\ \tau_{xy} \end{Bmatrix}^k = \begin{bmatrix} \bar{Q}_{11} & \bar{Q}_{12} & \bar{Q}_{16} \\ \bar{Q}_{12} & \bar{Q}_{22} & \bar{Q}_{26} \\ \bar{Q}_{16} & \bar{Q}_{26} & \bar{Q}_{66} \end{bmatrix}^k \begin{Bmatrix} \overset{\circ}{\varepsilon}_x \\ \overset{\circ}{\varepsilon}_y \\ \overset{\circ}{\gamma}_{xy} \end{Bmatrix}$$

16.10.2 Confirm expression (16.13) for the laminate inplane stiffness $[A]$.

Recall:

$$\{\sigma\}^k = [\bar{Q}]^k(\{\varepsilon^\circ\} + z\{\kappa\})$$

$$\{N\} = \int_{-h}^{h} \{\sigma\}^k dz$$

Therefore,

$$[A] \equiv \int_{-h}^{h} [\bar{Q}]\{\varepsilon^\circ\}dz$$

$$[A] \equiv \{\varepsilon^\circ\} \int_{-h}^{h} [\bar{Q}]dz = \{\varepsilon^\circ\} \sum_{k=1}^{N} [\bar{Q}]^k(z_k - z_{k-1})$$

The integral is replaced by a summation because $[\bar{Q}]^k$ is constant in each layer.

16.10.3 Confirm the expression for the bending–stretching coupling $[B]$ in Eq. (16.13).

Solution

$$[B] \equiv \int_{-h}^{h} [\bar{Q}]zdz$$

$$[B] = \sum_{k=1}^{N} [\bar{Q}]^k \frac{(z_k^2 - z_{k-1}^2)}{2} = \frac{1}{2}\sum_{k=1}^{N} [\bar{Q}]^k(z_k^2 - z_{k-1}^2)$$

16.10.4 Confirm the expression for the bending stiffness $[D]$ in Eq. (16.16).

$$\{\sigma\}^k = [\bar{Q}]^k(\{\varepsilon^\circ\} + z\{\kappa\})$$

$$\{M\} = \int_{-h}^{h} \{\sigma\}^k z dz$$

$$[D] = \int_{-h}^{h} [\bar{Q}] z^2 dz = \frac{1}{3}\sum_{k=1}^{N} [\bar{Q}]^k (z_k^3 - z_{k-1}^3)$$

References

Herakovich, C. T. (1998). *Mechanics of fibrous composites*. New York: Wiley.
Herakovich, C. T. (2012). *Mechanics of composites: A historical review*. Amsterdam: Elsevier, Mechanics Research Communications.
Kirchhoff, G. (1850). J. Math (Crelle) bd. 40.

Appendix

A.1 Trigonometric Identities

$$\sin^2\theta + \cos^2\theta = 1$$
$$\sin 2\theta = 2\sin\theta\cos\theta$$
$$\cos 2\theta = \cos^2\theta - \sin^2\theta$$
$$\sin^2\theta = \frac{1 - \cos 2\theta}{2}$$
$$\cos^2\theta = \frac{1 + \cos 2\theta}{2}$$

A.2 Derivatives of Trigonometric Functions

$$\frac{d}{d\theta}\sin\theta = \cos\theta$$
$$\frac{d}{d\theta}\cos\theta = -\sin\theta$$
$$\frac{d}{d\theta}\sin^2\theta = 2\sin\theta\cos\theta = \sin 2\theta$$

C.T. Herakovich, *A Concise Introduction to Elastic Solids*,
DOI 10.1007/978-3-319-45602-7

A.3 Moment of Inertia

$$I_x = \int_A y^2 dA$$

$$I_y = \int_A x^2 dA$$

$$J = \int_A r^2 dA$$

A.4 Partial Derivatives

For a function, f, of two variables, say x and y, we write $f = f(x, y)$. The total differential df is then:

$$df(x,y) = \frac{\partial f(x,y)}{\partial x} dx + \frac{\partial f(x,y)}{\partial y} dy$$

$\frac{\partial f}{\partial x}$ and $\frac{\partial f}{\partial y}$ are the partial derivatives of $f = f(x, y)$. The partial derivative with respect to one variable represents the change due to that variable while holding the other variable constant.

A.5 Biharmonic Operator ∇^4

The two-dimensional biharmonic operator in the x–y plane is defined as:

$$\nabla^4 \equiv \frac{\partial^4}{\partial x^4} + \frac{\partial^4}{\partial y^4} + 2\frac{\partial^4}{\partial x^2 \partial y^2}$$

A.6 Matrix Notation

Matrix notation is a shorthand way for expressing more expansive quantities. Examples follow.

One-dimensional array (also called a vector) in three-dimensional space (i = 1, 2, 3). A_1, A_2 and A_3 are the *components* of the vector:

$$\{A\} \equiv A_i \equiv \begin{Bmatrix} A_1 \\ A_2 \\ A_3 \end{Bmatrix}$$

Two-dimensional array (i, j = 1, 2, 3):

$$[A] \equiv A_{ij} \equiv \begin{bmatrix} A_{11} & A_{12} & A_{13} \\ A_{21} & A_{22} & A_{23} \\ A_{31} & A_{32} & A_{33} \end{bmatrix}$$

The transpose A_{ij}^T of a matrix a_{ij} is obtained by interchanging the rows and columns of A_{ij}. Thus,

$$A_{ij}^T \equiv A_{ji}$$

A.7 Matrix Determinate

The determinate Δ of a second-order matrix is defined as:

$$\Delta \equiv \begin{vmatrix} A_{11} & A_{12} \\ A_{21} & A_{22} \end{vmatrix} = A_{11}A_{22} - A_{12}A_{21}$$

The determinate Δ of a third-order matrix is defined as:

$$\Delta \equiv \begin{vmatrix} A_{11} & A_{12} & A_{13} \\ A_{21} & A_{22} & A_{23} \\ A_{31} & A_{32} & A_{33} \end{vmatrix}$$

$$= A_{11}A_{22}A_{33} + A_{12}A_{23}A_{31} + A_{13}A_{21}A_{32} - A_{13}A_{22}A_{31} - A_{11}A_{23}A_{32} - A_{12}A_{21}A_{33}$$

A.8 Matrix Inverse

Matrix inverse $[A]^{-1}$ of a matrix $[A]$ is defined as:

$$[A]^{-1} = \frac{1}{|\Delta|} \begin{bmatrix} a_{11} & a_{12} & a_{13} \\ a_{21} & a_{22} & a_{23} \\ a_{31} & a_{32} & a_{33} \end{bmatrix}^T = \frac{1}{|\Delta|} \begin{bmatrix} a_{11} & a_{21} & a_{31} \\ a_{12} & a_{22} & a_{32} \\ a_{13} & a_{23} & a_{33} \end{bmatrix}$$

where the a_{ij} are the cofactors of $[A]$ determined by suppressing ith row and jth column from $[A]$ and forming the determinate of the remaining matrix (called the minor of A_{ij}) and then multiplying the minor by $(-1)^{i+j}$.

With the result for a third-order matrix,

$$[A]^{-1}=\frac{1}{|\Delta|}\begin{bmatrix} (A_{22}A_{33}-A_{23}A_{32}) & -(A_{12}A_{33}-A_{13}A_{32}) & (A_{12}A_{23}-A_{13}A_{22}) \\ -(A_{21}A_{33}-A_{23}A_{31}) & (A_{11}A_{33}-A_{13}A_{31}) & -(A_{11}A_{23}-A_{13}A_{21}) \\ (A_{21}A_{32}-A_{22}A_{31}) & -(A_{11}A_{32}-A_{12}A_{31}) & (A_{11}A_{22}-A_{12}A_{21}) \end{bmatrix}$$

Kronecker delta $[\delta]$ is defined as:

$$[\delta] \equiv \begin{bmatrix} 1 & 0 & 0 \\ 0 & 1 & 0 \\ 0 & 0 & 1 \end{bmatrix}$$

Thus,

$$[A][A]^{-1}=[\delta]$$

A.9 Tensors

Tensors are mathematical representations of physical quantities. A very important property of tensors is that they obey certain laws of transformation between coordinate systems. Tensors are written in index notation; the index notation can be interpreted much like matrix notation. A tensor quantity f_i is said to have one live subscript, i. If an index i is repeated, e.g., f_{ii}, the index is called a dummy index and summation is implied. Thus, in *three-dimensional space* $i = 1, 2, 3$:

$$f_{ii} = f_{11} + f_{22} + f_{33}$$

Different levels of tensors are defined as follows:

Scalar—0 live index—f or f_{ii} (density is an example of a scalar; it is independent of direction).

Vector—1 live index—f_i (three components: force is an example; it may have components in thee directions).

$$F_i = \begin{bmatrix} F_1 \\ F_2 \\ F_3 \end{bmatrix}$$

For $F_3 = 0$, graphical representation of the force and its components in the x_1–x_2 plane is as shown below.

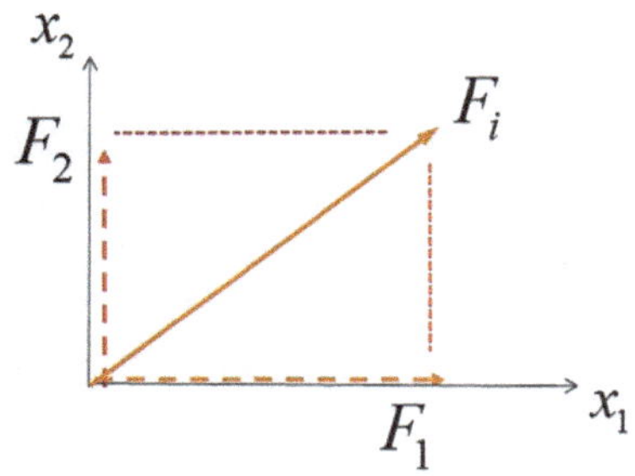

Second-Order Tensor—2 live index σ_{ij} (stress and strain are examples)

$$\sigma_{ij} = \begin{bmatrix} \sigma_{11} & \sigma_{12} & \sigma_{13} \\ \sigma_{21} & \sigma_{22} & \sigma_{23} \\ \sigma_{31} & \sigma_{32} & \sigma_{33} \end{bmatrix}$$

A.10 Tensor Transformations

It is often desired to determine the planes of maximum and/or minimum values of a physical quantity. The tensor transformation equations are ideal for determining quantities as a function of the orientation of the coordinate system. Stress and strain are good examples of tensor quantities that obey the tensor transformation relationships described below.

Tensors have components that change from one coordinate system to another (e.g., unprimed coordinates to primed coordinates as in the figure below) according to the so-called transformation equations. We will be concerned only with rectangular Cartesian coordinates and hence *Cartesian tensors.* Tensors usually are written using index notation, with the order of a tensor indicated by the number of live (non-repeated) subscripts. If a subscript is repeated, summation over the range of that subscript is implied, unless otherwise indicated. For rectangular Cartesian coordinates, the subscripts have a range of 1, 2, and 3 in three-dimensional space and a range of 1 and 2 in two-dimensional space.

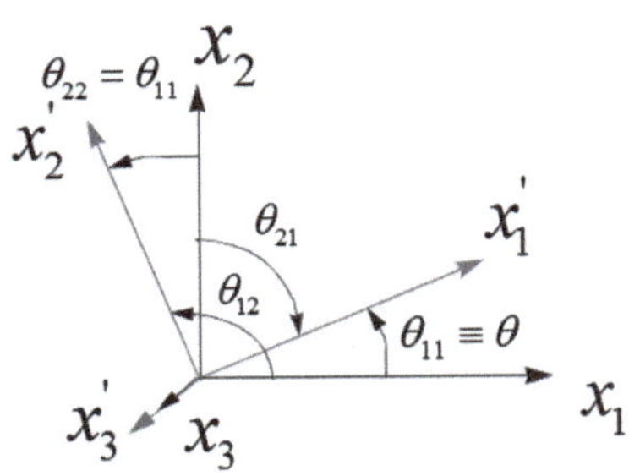

The equations for transforming a tensor quantity from one rectangular coordinate system to another are written in terms of the direction cosines, a_{ij}, (i, j = 1, 2, 3) of the angles measured from the unprimed axes, x_i, to the primed axes, x'_i, (e.g., $a_{12} = \cos\theta_{12}$ in the figure). *In this book, we shall use the convention that the first subscript (i) of* a_{ij} *corresponds to the initial unprimed axes and the second subscript (j) corresponds to the final primed axes.* (The reader is forewarned that some authors use the opposite convention for a_{ij}, i.e., the first subscript corresponds to the final, primed axes.)

Using trigonometry, we summarize the direction cosines for rotation θ about the 3-axis as:

$a_{11} = \cos\theta_{11} = \cos\theta$	$a_{12} = \cos\theta_{12} = -\sin\theta$	$a_{13} = \cos 90 = 0$
$a_{21} = \cos\theta_{21} = \sin\theta$	$a_{22} = \cos\theta_{22} = \cos\theta$	$a_{23} = \cos 90 = 0$
$a_{31} = \cos 90 = 0$	$a_{32} = \cos 90 = 0$	$a_{33} = \cos 0 = 1$

In matrix form, the transformation coefficients a_{ij} are:

$$a_{ij} = \begin{bmatrix} a_{11} & a_{12} & a_{13} \\ a_{21} & a_{22} & a_{23} \\ a_{31} & a_{32} & a_{33} \end{bmatrix} = \begin{bmatrix} \cos\theta & -\sin\theta & 0 \\ \sin\theta & \cos\theta & 0 \\ 0 & 0 & 1 \end{bmatrix}$$

In a continuing effort to simplify the writing of equations, it is common to define $m \equiv \cos\theta$ and $n \equiv \sin\theta$ with the transformation coefficients for rotation about the 3-axis then being:

$$a_{ij} = \begin{bmatrix} m & -n & 0 \\ n & m & 0 \\ 0 & 0 & 1 \end{bmatrix}$$

A.11 Vector Transformation

The transformation equations for a first-order tensor F_i in three-dimensional space are:

$$F'_i = a_{ji}F_j$$
$$F'_i = a_{1i}F_1 + a_{2i}F_2 + a_{3i}F_3$$

And, the individual components are:

$$F'_1 = a_{11}F_1 + a_{21}F_2 + a_{31}F_3$$
$$F'_2 = a_{12}F_1 + a_{22}F_2 + a_{32}F_3$$
$$F'_3 = a_{13}F_1 + a_{23}F_2 + a_{33}F_3$$

A.12 Second-Order Tensor Transformation

The transformation equations for a second-order tensor σ_{ij} are:

$$\sigma'_{ij} = a_{ki}a_{lj}\sigma_{kl}$$

In two-dimensional space $(i,j = 1,2)$, this becomes:

$$\sigma'_{ij} = a_{1i}a_{1j}\sigma_{11} + a_{1i}a_{2j}\sigma_{12} + a_{2i}a_{1j}\sigma_{21} + a_{2i}a_{2j}\sigma_{22}$$

And, in three-dimensional space with $(i,j = 1,2,3)$, the transformation equations are:

$$\begin{aligned}\sigma'_{ij} = {} & a_{1i}a_{1j}\sigma_{11} + a_{1i}a_{2j}\sigma_{12} + a_{li}a_{3j}\sigma_{13} + a_{2i}a_{1j}\sigma_{21} + a_{2i}a_{2j}\sigma_{22} \\ & + a_{2i}a_{3j}\sigma_{23} + a_{3i}a_{1j}\sigma_{31} + a_{3i}a_{2j}\sigma_{32} + a_{3i}a_{3j}\sigma_{33}\end{aligned}$$

Index

C.T. Herakovich, *A Concise Introduction tc Elastic Solids*,
DOI 10.1007/978-3-319-45602-7

FSC
www.fsc.org
MIX
Papier | Fördert gute Waldnutzung
FSC® C083411

Kolibri 360 GmbH
Ferdinand-Jühlke-Straße 7
99095 Erfurt, Deutschland
produktsicherheit@kolibri360.de